OPTICAL PHYSICS AND ENGINEERING

Editor: **William L. Wolfe**
Optical Sciences Center, University of Arizona, Tucson, Arizona

M. A. Bramson
Infrared Radiation: A Handbook for Applications

Sol Nudelman and S. S. Mitra, Editors
Optical Properties of Solids

S. S. Mitra and Sol Nudelman, Editors
Far-Infrared Properties of Solids

Lucien M. Biberman and Sol Nudelman, Editors
Photoelectronic Imaging Devices
    Volume 1: Physical Processes and Methods of Analysis
    Volume 2: Devices and Their Evaluation

A. M. Ratner
Spectral, Spatial, and Temporal Properties of Lasers

Lucien M. Biberman, Editor
Perception of Displayed Information

W. B. Allan
Fibre Optics: Theory and Practice

Albert Rose
Vision: Human and Electronic

J. M. Lloyd
Thermal Imaging Systems

Winston E. Kock
Engineering Applications of Lasers and Holography

Shashanka S. Mitra and Bernard Bendow, Editors
Optical Properties of Highly Transparent Solids

M. S. Sodha and A. K. Ghatak
Inhomogeneous Optical Waveguides

A Continuation Order Plan is available for this series. A continuation order will bring delivery of each new volume immediately upon publication. Volumes are billed only upon actual shipment. For further information please contact the publisher.

Series

Inhomogeneo

Optical Waveg

Inhomogeneous Optical Waveguides

M. S. Sodha and A. K. Ghatak

Indian Institute of Technology
New Delhi, India

PLENUM PRESS · NEW YORK AND LONDON

Library of Congress Cataloging in Publication Data

Sodha, M   S   1932-
  Inhomogeneous optical waveguides.

  (Optical physics and engineering)
  Bibliography: p.
  Includes index.
  1. Optical wave guides. 2. Optical communications. I. Ghatak, Ajoy K., 1939-
      joint author. II. Title.
TK5103.59.S6   621.38'0414   76-49643
ISBN 0-306-30916-5

© 1977 Plenum Press, New York
A Division of Plenum Publishing Corporation
227 West 17th Street, New York, N.Y. 10011

Printed in the United States of America

Preface

The propagation of electromagnetic waves in "square-law" media, i.e., media characterized by a quadratic spatial variation of the dielectric constant, has been a favorite subject of investigation in electromagnetic theory. However, with the recent fabrication of glass fibers with a quadratic radial variation of the dielectric constant and the application of such fibers to optical imaging and communications, this subject has also assumed practical importance. Comparison of experimental results on propagation, resolution, and pulse distortion in such inhomogeneous waveguides with theory has put the field on a sound base and spurred further work.

The present book aims at presenting a unified view of important aspects of our knowledge of inhomogeneous optical waveguides. A brief discussion of homogeneous dielectric waveguides is unavoidable, since it forms a basis for the appreciation of inhomogeneous waveguides. A short course based on some chapters of this book was offered to graduate students at IIT Delhi and was well received. We consider that despite the unavoidable mathematical nature of the present book, the comparison of experimental results with theory throughout and the description of fabrication technology (Appendixes A and B) should make its appeal universal.

The authors are grateful to Dr. K. Thyagarajan for writing most of Chapter 9 and to their colleagues Dr. I. C. Goyal, Dr. B. P. Pal, and Dr. A. Kumar for helpful discussions and valuable criticisms. The authors also express grateful thanks to the late Prof. L. Kraus and Dr. P. Herczfeld of Drexel University, Philadelphia, for stimulating discussions during the authors' visit to Drexel and the latters' visit to IITD; the National Science Foundation, U.S.A., supported these visits.

The writing of this book constitutes an activity in The Laser Applications Program of the Institute under the National Science and Technology Plan. The financial support of the National Science Foundation for the original work by the authors reported in this book is gratefully acknowledged.

M.S.S.
A.K.G.

Contents

CHAPTER 1

Introduction

One of the most exciting applications of lasers lies in the field of communications, i.e., the use of laser beams to carry information in a way similar to that employing radiowaves, microwaves, and millimeter waves. The chief advantage of a laser communication system lies in its huge information-carrying capacity, a result of the high frequency of the carrier. However, unlike other carriers, laser beams propagating through the atmosphere cannot be used for communications (for distances greater than a few kilometers) because the scattering by rain, snow, dust, clouds, and so on, leads to large attenuation; further, unacceptable levels of noise are introduced by the turbulence of the air. In addition to attenuation by the atmospheric aerosols, the beam also diverges because of diffraction, resulting in a loss of intensity in the central portion of the beam. Thus there is a necessity for some kind of device which will guide the propagation of the beam so that the power of the beam is confined in the transverse direction; by care in manufacture the attenuation and noise is also minimized. Such devices are commonly known as waveguides.

The earliest known form of an optical waveguide consisted of a transparent cylindrical rod surrounded by a material having a refractive index lower than the material of the cylinder; the guidance of the optical beam takes place by multiple total internal reflections from the surface of the cylinder. This phenomenon of guidance by multiple total internal reflections was demonstrated by John Tyndall at a meeting of the Royal Society, London, in 1870. In this demonstration Tyndall showed that light travels along the curved path of a stream of water emanating from an illuminated vessel.

The principle of guidance by multiple reflections has been exploited in the field of fiber optics; for an introduction to the subject the reader is referred to the excellent text by Kapany (1967). This book discusses the applications of fibers for transmission and scanning of images.

In the early applications of fiber optics, an approach based on geometrical optics was adequate for an understanding of the basic phenomena and applications. For later applications (as the diameter of the fiber approached the wavelength of light) a treatment based on the solution of Maxwell's equations became necessary. A good review of the work on electromagnetic wave propagation in such waveguides (based on Maxwell's equations) has been presented by Kapany and Burke (1972) and Marcuse (1972, 1974). These books also include an analysis of propagation through a dielectric slab coated by material having a lower dielectric constant; such slab waveguides are of considerable importance in the growing field of integrated optics.

Apart from imaging applications, optical waveguides have recently been used for optical communications; it is visualized that this will be the major application in years to come. Low loss, minimum signal distortion, and the ability to pass several laser beams simultaneously (without mixing) are the primary requirements for the suitability of optical waveguides for this purpose.

Recently, circular optical waveguides characterized by the spatially varying dielectric constant,

$$K(r) = K_0 - K_2 r^2 \qquad (1.1)$$

have been fabricated (known as SELFOC and GRIN fibers); planar inhomogeneous waveguides with dielectric constant varying along one direction have also been fabricated.

Table 1.1 outlines a comparison of the main features of inhomogeneous waveguides and dielectric-clad homogeneous waveguides.

Inhomogeneous optical waveguides (slab and cylindrical types) constitute the subject of the present book. A minimum analysis of the conventional (dielectric-clad homogeneous) waveguides has also been presented to form a basis for the appreciation of inhomogeneous waveguides.

Chapter 2 presents a general modal analysis of slab waveguides and discusses in detail the modes in a dielectric-clad homogeneous waveguide; the corresponding discussion for inhomogeneous slab waveguides is given in Chapter 3. Chapters 4 and 5 present analyses of electromagnetic propagation in dielectric-clad homogeneous circular waveguide and inhomogeneous waveguides, characterized by Eq. (1.1); these analyses are based on the scalar wave equation; the vector theory for the inhomogeneous guides is

Table 1.1. *Comparison of Inhomogeneous and Homogeneous Waveguides*

Feature	Inhomogeneous waveguide	Homogeneous waveguide
Structure	$K = K_0 - K_2 r^2$ for $r < a$ $K = K_1$ for $r > a$	$K = K_1$ for $r < a$ $K = K_2$ for $r > a$
Waveguiding action	The light beam heading for the interface is always pulled back to the central axis because of the refractive index gradient	Multiple total internal reflections at the interface
Path of rays	Sinusoidal path of rays in inhomogeneous waveguides	Path of rays in homogeneous waveguides
Pulse propagation	High-frequency light pulse can be passed without distortion of waveform and phase	High-frequency light pulse cannot be passed without distortion of waveform and phase
Transmission loss	Very low	Higher
Lens action	Yes	No

given in Chapter 6. Propagation and distortion of laser signals in both homogeneous and inhomogeneous optical waveguides is the subject matter of Chapter 7; it is seen that the inhomogeneous optical waveguide is more advantageous because the group velocity of all the modes in such a waveguide is, to a first approximation, constant. The role of imperfections in waveguides is discussed in Chapter 8. Since inhomogeneous waveguides can also be used as lenses, an adequate treatment of geometrical ray tracing and image aberrations is given in Chapter 9.

CHAPTER 2

Modal Analysis of Slab Waveguides

2.1. Introduction

It is well known that a beam of light propagating in a homogeneous medium suffers divergence as a result of diffraction, and is thus unable to traverse appreciable distances without loss of intensity in the central portion of the beam. For many applications it is essential to avoid this divergence, or in other words guide the propagation of the beam (or the wave) so that the power of the beam is confined in the transverse direction; devices for guiding light waves are known as optical waveguides. In the simplest waveguides the power of a beam (propagating along the z direction) is confined between two parallel planes, say $x = \pm a$; such waveguides are known as slab waveguides and consist of a layer of material with high refractive index $(-a < x < a)$ sandwiched between two layers of a material with a slightly lower refractive index (see Fig. 2.1). The slab waveguide owes its effectiveness to the fact that most of the diverging rays (not coincident with the z axis) are incident on the boundary at an angle exceeding the critical angle for the interface and hence are subject to total internal reflection (without loss of power) into the beam. The fabrication and performance of such waveguides, known as dielectric-clad planar waveguides, is discussed in Appendix A. Such waveguides now form an essential component of most present-day integrated circuits. Although the purpose of this book is to present an account

5

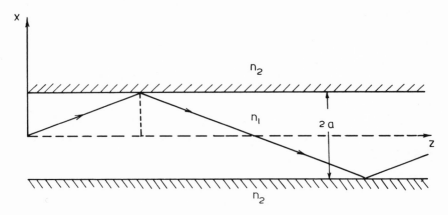

Fig. 2.1. In a slab waveguide the refractive index has a step discontinuity at $x = \pm a$. The region $|x| < a$ is known as the core and the region $|x| > a$ is known as the cladding. The core refractive index n_1 is greater than the cladding refractive index n_2. The medium is assumed to be of infinite extent in the y direction. If a ray is incident on the core–cladding interface making an angle greater than the critical angle then the ray will undergo total internal reflections and will be trapped inside the core.

of inhomogeneous waveguides in which the guidance of waves is achieved by a continuous change of refractive index in the medium as compared to the step change in the refractive index mentioned above, we have devoted this chapter and part of Appendix A to a discussion of the latter. This has been necessitated by the fact that an appreciation of inhomogeneous waveguides is not possible without comparison with a dielectric-clad waveguide.

In order to understand the performance of the waveguides it is necessary to obtain the appropriate solutions of Maxwell's equations. In Sec. 2.2, on the basis of general theoretical considerations, we derive wave equations which determine the modes* and are valid for arbitrary x dependence of the dielectric constant. In Sec. 2.3 we have analyzed the waveguiding action in a simple planar waveguide having a step variation of the dielectric constant

---

*It is seen that if the distribution of the electric field (amplitude and phase) in a transverse cross section of the beam is arbitrary, the nature of the distribution changes as the beam propagates in the waveguide. However, for certain specific distributions of the electric vector in the cross section, the nature of the distribution remains unchanged with propagation; such distributions of electric vector in the plane of the transverse cross section are commonly referred to as modes. Different modes propagate with different group velocities and suffer different attenuations. Any arbitrary distribution of electric field in the cross section of a beam at $z = 0$ can be, in general, expressed (to a good approximation) as a resultant of different modes with different amplitudes and phases which propagate with different propagation parameters. The resulting distribution in a cross section at z may be obtained by summing over different modes.

characterized by

$$K(x) = K_1 \qquad \text{for } |x| < a$$
$$= K_2 \qquad \text{for } |x| > a$$

This is followed by a study of asymmetric* planar waveguides, taking into account the effect of absorption. In the next chapter we will discuss the waveguiding action in planar waveguides with continuous dielectric constant variation. In Chapter 6, we will discuss various approximate methods used in solving the wave equation.

2.2. General Theoretical Considerations

We start with Maxwell's equations

$$\nabla \times \mathscr{E} + \frac{\partial \mathbf{B}}{\partial t} = 0 \tag{2.1}$$

$$\nabla \times \mathscr{H} = \mathbf{J} + \frac{\partial \mathbf{D}}{\partial t} \tag{2.2}$$

$$\nabla \cdot \mathbf{D} = \rho \tag{2.3}$$

$$\nabla \cdot \mathbf{B} = 0 \tag{2.4}$$

where $\mathscr{E}$, $\mathbf{D}$, $\mathbf{B}$, and $\mathscr{H}$ represent the electric field, electric displacement, magnetic induction, and magnetic intensity, respectively; $\mathbf{J}$ and ρ denote the conduction current density and charge density, respectively. The MKS system of units has been used in writing Eqs. (2.1)–(2.4). For isotropic, linear, nonconducting, and nonmagnetic material we further have

$$\mathbf{J} = 0 \tag{2.5}$$

$$\rho = 0 \tag{2.6}$$

$$\mathbf{B} = \mu_0 \mathscr{H} \tag{2.7}$$

$$\mathbf{D} = \varepsilon \mathscr{E} \tag{2.8}$$

where μ_0 is the magnetic permeability of free space ($4\pi \times 10^{-7}$ N/A^2) and ε the dielectric permittivity. Often it is more convenient to work with the dimensionless dielectric constant, K, defined by the relation

$$K = \varepsilon/\varepsilon_0 \tag{2.9}$$

---

*In asymmetric waveguides electromagnetic waves propagate in a medium sandwiched between two different media.

where ε_0 is the permittivity of free space and is equal to 8.854×10^{-12} $C^2/N \ m^2$. In this and the following chapter we will assume ε (and therefore K) to be a function of the x coordinate only.

Taking the curl of Eq. (2.1) and using Eq. (2.7) we obtain [using Eqs. (2.2), (2.6), and (2.8)]

$$\mathbf{\nabla} \times (\mathbf{\nabla} \times \mathcal{E}) = -\mu_0 \frac{\partial}{\partial t} \mathbf{\nabla} \times \mathcal{H}$$

$$= -\mu_0 \frac{\partial}{\partial t} \left(\varepsilon \frac{\partial \mathcal{E}}{\partial t} \right)$$

$$= -\varepsilon_0 \mu_0 K \frac{\partial^2 \mathcal{E}}{\partial t^2}$$

Thus

$$-\nabla^2 \mathcal{E} + \mathbf{\nabla}(\mathbf{\nabla} \cdot \mathcal{E}) = -\frac{1}{c^2} K \frac{\partial^2 \mathcal{E}}{\partial t^2} \tag{2.9a}$$

where $c = (\varepsilon_0 \mu_0)^{-1/2}$ is the speed of light in free space and use has been made of the vector identity

$$\mathbf{\nabla} \times (\mathbf{\nabla} \times \mathcal{E}) = -\nabla^2 \mathcal{E} + \mathbf{\nabla}(\mathbf{\nabla} \cdot \mathcal{E}) \tag{2.10}$$

From Eqs. (2.3), (2.6), and (2.8) we obtain

$$0 = \mathbf{\nabla} \cdot \mathbf{D} = \mathbf{\nabla} \cdot (\varepsilon \mathcal{E}) = \varepsilon \mathbf{\nabla} \cdot \mathcal{E} + \mathcal{E} \cdot \mathbf{\nabla} \varepsilon \tag{2.11}$$

or

$$\mathbf{\nabla} \cdot \mathcal{E} = -\frac{1}{\varepsilon}(\mathcal{E} \cdot \mathbf{\nabla} \varepsilon) = -\frac{1}{K}(\mathcal{E} \cdot \mathbf{\nabla} K) \tag{2.12}$$

On substitution in Eq. (2.9) and assuming a time dependence of the form $\exp(i\omega t)$ we finally obtain

$$\nabla^2 \mathcal{E} + \mathbf{\nabla} \left(\mathcal{E} \cdot \frac{1}{K} \mathbf{\nabla} K \right) + \frac{K(x)\omega^2}{c^2} \mathcal{E} = 0 \tag{2.13}$$

In a similar manner we obtain

$$\nabla^2 \mathcal{H} + \frac{1}{K}(\mathbf{\nabla} K) \times (\mathbf{\nabla} \times \mathcal{H}) + \frac{K(x)\omega^2}{c^2} \mathcal{H} = 0 \tag{2.14}$$

With practical systems very little error is introduced if we neglect the second term depending on $(1/K)\mathbf{\nabla}K$; in general, this is justified (as we will show later) when the variation of the dielectric constant is small in distances of the order of the wavelength (this is almost always true, particularly for optical beams). However, for planar waveguides the dielectric constant is constant in each region, and therefore as long as we solve the wave equation separately in each region, the term depending on ∇K vanishes identically. In the next chapter we will assume a continuous variation of the dielectric constant and will explicitly evaluate the effect of ∇K on the propagation constant.

The x components of Eqs. (2.13) and (2.14) are given by

$$\nabla^2 \mathscr{E}_x + \frac{\partial}{\partial x}\left[\frac{1}{K}\frac{dK}{dx}\mathscr{E}_x\right] + \frac{\omega^2}{c^2}K(x)\mathscr{E}_x = 0 \tag{2.15}$$

and

$$\nabla^2 \mathscr{H}_x + (\omega^2/c^2)K(x)\mathscr{H}_x = 0 \tag{2.16}$$

In Eqs. (2.15) and (2.16) we have only the x component of $\mathscr{E}$ and $\mathscr{H}$, respectively, because K depends only on x. The y and z components of Eqs. (2.13) and (2.14) are inconvenient to work with because they involve more than one component of $\mathscr{E}$ and $\mathscr{H}$. Hence for further analysis we will be relying on Eqs. (2.15) and (2.16); the other components of $\mathscr{E}$ and $\mathscr{H}$, corresponding to any solution for $\mathscr{E}_x$ and $\mathscr{H}_x$, have been written down from Maxwell's equations in later analysis.

The y and z dependence of the field components $\mathscr{E}_x$ and $\mathscr{H}_x$ are of the form

$$\exp[-i(\beta z + \gamma y)] \tag{2.17}$$

this can be seen if we try to solve either Eq. (2.15) or (2.16) by the method of separation of variables. Thus if

$$\mathscr{E}_x = X(x)Y(y)Z(z) \tag{2.18}$$

is substituted in Eq. (2.15) and the resulting equation is divided by XYZ, we obtain

$$\left\{\frac{1}{X}\left[\frac{d^2X}{dx^2} + \frac{d}{dx}\left(\frac{1}{K}\frac{dK}{dx}X\right)\right] + \frac{\omega^2}{c^2}K(x)\right\}$$

$$+ \left(\frac{1}{Y}\frac{d^2Y}{dy^2}\right) + \left(\frac{1}{Z}\frac{d^2Z}{dz^2}\right) = 0 \tag{2.19}$$

The first term (inside the square brackets) is a function of x only, the second term is a function of y only, and the third term is a function of z only; consequently, we must set each term equal to a constant. We write

$$\frac{1}{Y}\frac{d^2Y}{dy^2} = -\gamma^2 \tag{2.20}$$

A positive quantity on the RHS (right-hand side) of (2.20) would imply fields which would go to infinity either in the $+y$ or $-y$ direction and hence is inadmissible. The solution of the above equation is of the form $\exp(\pm i\gamma y)$. Similarly,

$$\frac{1}{Z}\frac{d^2Z}{dz^2}=-\beta^2 \tag{2.21}$$

the solution of which is of the form of $\exp(\pm i\beta z)$.

Thus, substituting for the y and z dependence of $\mathscr{E}_x$ and $\mathscr{H}_x$ from Eq. (2.17) in Eqs. (2.15) and (2.16), one obtains

$$\frac{d^2E_x}{dx^2}+\frac{d}{dx}\left[\frac{1}{K}\frac{dK}{dx}E_x\right]+\left[\frac{\omega^2}{c^2}K(x)-\beta^2-\gamma^2\right]E_x=0 \tag{2.22a}$$

and

$$\frac{d^2H_x}{dx^2}+\left[\frac{\omega^2}{c^2}K(x)-\beta^2-\gamma^2\right]H_x=0 \tag{2.22b}$$

where E_x and H_x are defined through the relations

$$\mathscr{E}_x=E_x\exp[i(\omega t-\beta z-\gamma y)] \tag{2.23a}$$

and

$$\mathscr{H}_x=H_x\exp[i(\omega t-\beta z-\gamma y)] \tag{2.23b}$$

Equation (2.22a) can be transformed to

$$\frac{d^2\phi}{dx^2}+\left[\frac{\omega^2}{c^2}K(x)-\frac{3}{4}\left(\frac{1}{K}\frac{dK}{dx}\right)^2+\frac{1}{2K}\frac{d^2K}{dx^2}-\beta^2-\gamma^2\right]\phi=0 \tag{2.24a}$$

where

$$\phi=[K(x)]^{1/2}E_x \tag{2.24b}$$

We note that Eq. (2.24a) does not involve the first derivative of ϕ [cf. Eq. (2.22a)] and therefore can be solved for many dielectric constant profiles. Further, the term involving ∇K in the wave equation would make a negligible contribution if

$$\left|\frac{1}{K(x)}\left[\frac{1}{2K}\frac{d^2K}{dx^2}-\frac{3}{4}\left(\frac{1}{K}\frac{dK}{dx}\right)^2\right]\right|\ll\frac{\omega^2}{c^2}=\frac{4\pi^2}{\lambda_0^2} \tag{2.24c}$$

where λ_0 is the free-space wavelength. The above inequality qualitatively implies that the variation of the dielectric constant is small in distances of the order of a wavelength; this is satisfied for realistic optical waveguides.

For K depending on x only, it is convenient to consider modes in which either $E_x = 0$ or $H_x = 0$; so that only one differential equation is to be considered. In general, both the longitudinal components E_z and H_z will exist and all of the field components will (as a consequence of Maxwell's equations) have a y dependence of the form $\exp(-i\gamma y)$.

We will now consider two special cases: (1) $H_x = 0$ and (2) $E_x = 0$. Any distribution of fields can be expressed as a combination of the field distributions corresponding to the different modes obtained in the following two cases (Ghatak and Kraus, 1974).

Case 1. If $H_x = 0$ the other field components, in terms of E_x, are given by

$$E_z = -\frac{i\beta}{K} \frac{1}{\beta^2 + \gamma^2} \frac{\partial}{\partial x}[K(x)E_x] \tag{2.25}$$

$$E_y = -\frac{i\gamma}{K} \frac{1}{\beta^2 + \gamma^2} \frac{\partial}{\partial x}[K(x)E_x] \tag{2.26}$$

$$H_z = -\frac{\gamma\omega\varepsilon_0 K}{\beta^2 + \gamma^2}E_x \tag{2.27}$$

$$H_y = \frac{\beta\omega\varepsilon_0 K}{\beta^2 + \gamma^2}E_x \tag{2.28}$$

The above equations can easily be derived from Maxwell's equations [Eqs. (2.1)–(2.4)]. For example, Eqs. (2.1) and (2.2) lead to

$$-i\gamma E_z + i\beta E_y = -i\omega\mu_0 H_x \tag{2.29}$$

$$-i\beta E_x - \frac{\partial E_z}{\partial x} = -i\omega\mu_0 H_y \tag{2.30}$$

$$\frac{\partial E_y}{\partial x} + i\gamma E_x = -i\omega\mu_0 H_z \tag{2.31}$$

$$-i\gamma H_z + i\beta H_y = i\omega\varepsilon_0 K E_x \tag{2.32}$$

$$-i\beta H_x - \frac{\partial H_z}{\partial x} = i\omega\varepsilon_0 K E_y \tag{2.33}$$

$$\frac{\partial H_y}{\partial x} + i\gamma H_x = i\omega\varepsilon_0 K E_z \tag{2.34}$$

where we have used Eqs. (2.5)–(2.8) and (2.17). Further, using Eq. (2.17), Eq. (2.4) gives

$$\frac{\partial H_x}{\partial x} - i\gamma H_y - i\beta H_z = 0 \tag{2.35}$$

For $H_x = 0$, the above equation becomes

$$\gamma H_y + \beta H_z = 0 \tag{2.36}$$

Equations (2.32) and (2.36) would immediately give Eqs. (2.27) and (2.28). Equations (2.33) and (2.34) can then be used to give Eqs. (2.25) and (2.26).

Case 2. In a similar manner when $E_x = 0$, the other field components are, in terms of H_x, given by

$$H_z = -\frac{i\beta}{\beta^2 + \gamma^2}\frac{\partial H_x}{\partial x} \tag{2.37}$$

$$H_y = -\frac{i\gamma}{\beta^2 + \gamma^2}\frac{\partial H_x}{\partial x} \tag{2.38}$$

$$E_z = \frac{\gamma\omega\mu_0}{\gamma^2 + \beta^2}H_x \tag{2.39}$$

$$E_y = -\frac{\beta\omega\mu_0}{\gamma^2 + \beta^2}H_x \tag{2.40}$$

It can easily be seen that with proper orientation of the z axis, we may set $\gamma = 0$ without any loss of generality. For $\gamma = 0$ we can again consider the two cases:

(i) For the case $H_x = 0$ the longitudinal component of the magnetic field, $H_z = 0$ (see Eq. 2.27). Hence the magnetic field is transverse to the direction of propagation; such modes are known as transverse magnetic (or TM) modes.

(ii) When $E_x = 0$ the longitudinal component of the electric field, $E_z = 0$ (see Eq. 2.39). Hence the electric field is transverse to the direction of propagation; such modes are known as transverse electric (or TE) modes.

2.3. A Simple Planar Waveguide

Most of the salient characteristics of the modes in an optical waveguide can be understood by considering the simple planar waveguide in which the dielectric constant variation is of the form

$$\begin{aligned}K &= K_1 \qquad \text{for } |x| < a, \text{ region I} \\ &= K_2 \qquad \text{for } |x| > a, \text{ region II}\end{aligned} \tag{2.41}$$

The systems for which the above dielectric constant variation is valid are described in Appendix A. We will assume the waveguide to be of infinite extent in the x and y directions, and the z axis to coincide with the direction of propagation. We will assume the mode fields to be independent of y [i.e., $\gamma = 0$].

2.3.1. TE Modes

We first consider TE modes in which $E_z = 0$. For such a case, the only nonvanishing field components would be H_z, H_x, and E_y, and they would be related to each other through the following equations (see Sec. 2.2):

$$H_x = -\frac{i}{\omega\mu_0} \frac{\partial E_y}{\partial z} = -\frac{\beta}{\omega\mu_0} E_y \qquad (2.42)$$

$$H_z = \frac{i}{\omega\mu_0} \frac{\partial E_y}{\partial x} \qquad (2.43)$$

Since in each region the refractive index is a constant, the term proportional to ∇K would vanish and $\mathscr{E}_y$ would satisfy the following equation [see Eq. (2.13)]:

$$\frac{\partial^2 \mathscr{E}_y}{\partial x^2} + \frac{\partial^2 \mathscr{E}_y}{\partial z^2} = \frac{K}{c^2} \frac{\partial^2 \mathscr{E}_y}{\partial t^2} \qquad (2.44)$$

For waves propagating in the z direction, the z and t dependence of the field would be of the form

$$\exp[i(\omega t - \beta z)] \qquad (2.45)$$

As such, Eq. (2.44) assumes the form

$$\frac{\partial^2 E_y}{\partial x^2} + \left(\frac{\omega^2}{c^2} K_1 - \beta^2\right) E_y = 0 \qquad \text{for } |x| \leqslant a, \text{ region I} \qquad (2.46)$$

and

$$\frac{\partial^2 E_y}{\partial x^2} + \left(\frac{\omega^2}{c^2} K_2 - \beta^2\right) E_y = 0 \qquad \text{for } |x| \geqslant a, \text{ region II} \qquad (2.47)$$

where, as before,

$$\mathscr{E}_y = E_y \exp[i(\omega t - \beta z)]$$

Owing to the symmetry of the problem [i.e., $K(x) = K(-x)$], it is possible to have purely symmetric modes [i.e., $E_y(x) = E_y(-x)$] and purely antisymmetric modes* [i.e., $E_y(x) = -E_y(-x)$].

Symmetric TE Modes

For the symmetric mode [$E_y(x) = E_y(-x)$] the solutions of Eqs. (2.46) and (2.47) would be given by

$$E_y = A_s \cos px \qquad \text{for } |x| \le a \qquad (2.48)$$

$$= A_s \cos pa \exp[-\kappa(|x| - a)] \qquad \text{for } |x| \ge a \qquad (2.49)$$

where we have used the boundary condition† that E_y must be continuous at $|x| = a$. Further

$$p^2 = \frac{\omega^2}{c^2} K_1 - \beta^2 = \omega^2 \varepsilon_1 \mu_0 - \beta^2 \qquad (2.50)$$

and

$$\kappa^2 = \beta^2 - \frac{\omega^2}{c^2} K_2 = \beta^2 - \omega^2 \varepsilon_2 \mu_0 \qquad (2.51)$$

where $\varepsilon_1(= K_1 \varepsilon_0)$ and $\varepsilon_2(= K_2 \varepsilon_0)$ represent the permittivity of regions I and II, respectively. The subscript s on A in Eqs. (2.48) and (2.49) refers to the fact that we are considering symmetric modes. Since $\varepsilon_1 > \varepsilon_2$, both p^2 and κ^2 can be positive. When $\kappa^2 > 0$ the electric field outside the slab decays exponentially‡ and one obtains a guided mode; on the other hand for $\kappa^2 < 0$ the field is oscillatory outside the slab and one obtains a radiation mode.§ Thus the condition for guided propagation is

$$\kappa^2 > 0 \qquad (2.52)$$

*This is because of the fact that if $\psi(x)$ is a solution of the equation $(d^2\psi/dx^2) + F(x)\psi = 0$, and if $F(-x) = F(x)$ then $\psi(-x)$ is also a solution; this can be seen if x is replaced by $-x$. If $\psi(-x)$ is not a multiple of $\psi(x)$ (which will only be true if $\psi(x)$ is either symmetric or antisymmetric) then $\psi(-x)$ represents the other independent solution of the differential equation. Obviously any linear combination of $\psi(x)$ and $\psi(-x)$ is also a solution. We may choose these linear combinations in such a way that the solutions are either purely symmetric [$\psi(x) + \psi(-x)$] or purely antisymmetric [$\psi(x) - \psi(-x)$].

† The boundary conditions correspond to the fact that the longitudinal component of the fields should be continuous at the boundary. Thus, in the present problem, E_y, E_z, H_y, and H_z should be continuous at $x = \pm a$.

‡ The Poynting vector is also zero.

§ The Poynting vector is nonzero and hence energy is lost through the outer media.

The z component of the magnetic field, H_z, can easily be obtained through the relation

$$H_z = \frac{i}{\omega\mu_0} \frac{\partial E_y}{\partial x}$$

Thus

$$H_z = -\frac{ip}{\omega\mu_0} A_s \sin px \qquad \text{for } |x| \leq a \qquad (2.53)$$

$$= -\frac{i\kappa}{\omega\mu_0} A_s \cos pa \, \exp[-\kappa(|x|-a)] \qquad \text{for } x > a$$

$$\qquad\qquad\qquad\qquad\qquad\qquad\qquad\qquad\qquad\qquad (2.54)$$

$$= +\frac{i\kappa}{\omega\mu_0} A_s \cos pa \, \exp[-\kappa(|x|-a)] \qquad \text{for } x < -a$$

Continuity of H_z at $x = \pm a$ gives us

$$\mp\frac{ip}{\omega\mu_0} A_s \sin pa = \mp\frac{i\kappa}{\omega\mu_0} A_s \cos pa$$

or

$$\tan pa = \frac{\kappa}{p} \qquad (2.55)$$

For given values of ω, ε_1, ε_2, μ_0, and a the transcendental equation, Eq. (2.55), determines the propagation constants of the guided TE (symmetric) modes. The coefficient A_s can be related to the power P carried by the guided mode. In order to calculate the power P, we first calculate the Poynting vector $\mathbf{S}$, which is defined by

$$\mathbf{S} = \boldsymbol{\mathscr{E}} \times \boldsymbol{\mathscr{H}}$$

Now, if $f = f_0 \exp(i\omega t)$ and $g = g_0 \exp(i\omega t)$, where f_0 and g_0 may depend on other variables but not on time, then*

$$\overline{\mathrm{Re}\, f\, \mathrm{Re}\, g} = \tfrac{1}{2} \mathrm{Re}\, f^* g$$

where the bar indicates time averaging. Thus

$$\bar{S}_z = -\tfrac{1}{2} \mathrm{Re}\, (\mathscr{E}_y \mathscr{H}_x^*)$$

$$= -\tfrac{1}{2} \mathrm{Re}\, E_y \left(-\frac{i}{\omega\mu_0} \frac{\partial E_y}{\partial z} \right)^*$$

$$= \frac{\beta}{2\omega\mu_0} |E_y|^2 \qquad (2.56)$$

*$\mathrm{Re}\, f\, \mathrm{Re}\, g = \tfrac{1}{2} \mathrm{Re}\{fg + f^* g\}; \overline{fg} = 0.$

The power P carried by the guided mode would be given by

$$P = \frac{\beta}{2\omega\mu_0} \int_{-\infty}^{\infty} |E_y|^2 \, dx$$

$$= \frac{\beta}{\omega\mu_0} \int_0^{\infty} |E_y|^2 \, dx$$

$$= \frac{\beta}{\omega\mu_0} A_s^2 \left\{ \int_0^a \cos^2 px \, dx + \cos^2 pa \int_a^{\infty} \exp[-2\kappa(x-a)] \, dx \right\}$$

$$= \frac{\beta}{\omega\mu_0} A_s^2 \left(\frac{1}{2} a + \frac{1}{2p} \sin pa \cos pa + \frac{\cos^2 pa}{2\kappa} \right)$$

$$= \frac{\beta A_s^2}{\omega\mu_0} \left(\frac{1}{2} a + \frac{1}{2\kappa} \sin^2 pa + \frac{1}{2\kappa} \cos^2 pa \right)$$

where we have used Eq. (2.55). Thus

$$P = \frac{\beta}{2\omega\mu_0} \left(a + \frac{1}{\kappa} \right) A_s^2 \tag{2.57}$$

Antisymmetric TE Modes

The antisymmetric modes $[E_y(x) = -E_y(-x)]$ can be obtained in a manner similar to that described in earlier subsections. The field pattern is given by

$$E_y = A_a \sin px \qquad\qquad \text{for } |x| \leqslant a \tag{2.58}$$
$$= A_a \sin pa \exp[-\kappa(|x|-a)] \qquad \text{for } |x| \geqslant a$$

where the subscript a refers to the fact that we are considering antisymmetric modes. The longitudinal component of the magnetic field is given by

$$H_z = \frac{ip}{\omega\mu_0} A_a \cos px \qquad\qquad \text{for } |x| \leqslant a \tag{2.59}$$

$$= -\frac{i\kappa}{\omega\mu_0} A_a \sin pa \exp[-\kappa(|x|-a)] \qquad \text{for } x > a \tag{2.60}$$

$$= \frac{i\kappa}{\omega\mu_0} A_a \sin pa \exp[-\kappa(|x|-a)] \qquad \text{for } x < -a$$

The continuity of H_z at $x = \pm a$ leads to the following transcendental equation:

$$\tan pa = -p/\kappa \tag{2.61}$$

The solutions of Eqs. (2.55) and (2.61) will be discussed later.

2.3.2. TM Modes

We propose to carry out an analysis for TM modes (similar to that for TE modes) for which $H_z = 0$; then H_x and E_y would also vanish [see Eqs. (2.25)–(2.28)]. The only nonvanishing field components would be E_z, E_x, and H_y. Since the fields are assumed to be independent of y, i.e., $\gamma = 0$, the equation satisfied by H_y would be of the form

$$\frac{d^2 H_y}{dx^2} + \left(\frac{\omega^2}{c^2} K - \beta^2\right) H_y = 0 \tag{2.62}$$

with $K = K_1$ in region I and $K = K_2$ in region II. The time and z dependence has been assumed to be of the form given by Eq. (2.45):

$$\mathcal{H}_y = H_y \exp[i(\omega t - \beta z)]$$

Once H_y is known, the other nonvanishing field components are determined by

$$E_x = \frac{i}{\omega \varepsilon_0 K} \frac{\partial H_y}{\partial z} = \frac{\beta}{\omega \varepsilon} H_y \tag{2.63}$$

and

$$E_z = -\frac{i}{\omega \varepsilon_0 K} \frac{\partial H_y}{\partial x} \tag{2.64}$$

Continuity of the tangential components of the electric and magnetic fields at $x = \pm a$ would give us the complete field patterns and the transcendental equation which would determine the propagation constants.

The final expression for the field patterns for both TE and TM modes are summarized below.

Symmetric TE modes

$$\mathcal{E}_y = A_s \cos px \, \exp[i(\omega t - \beta z)]$$

$$\mathcal{H}_x = -\frac{\beta}{\omega \mu_0} \mathcal{E}_y$$

$$\mathcal{H}_z = -\frac{ip}{\omega \mu_0} \tan px \mathcal{E}_y \qquad \text{for } |x| \leq a \tag{2.65a}$$

$$\mathcal{E}_x = \mathcal{E}_z = 0$$

$$\mathcal{H}_y = 0$$

and

$$\mathscr{E}_y = A_s \cos pa \, \exp[-\kappa(|x|-a)]$$
$$\times \exp[i(\omega t - \beta z)]$$

$$\mathscr{H}_x = -\frac{\beta}{\omega\mu_0}\mathscr{E}_y$$

for $|x| \geqslant a$ (2.65b)

$$\mathscr{H}_z = -\frac{i\kappa|x|}{\omega\mu_0 x}\mathscr{E}_y$$

$$\mathscr{E}_x = \mathscr{E}_z = 0$$

$$\mathscr{H}_y = 0$$

Antisymmetric TE modes

$$\mathscr{E}_y = A_a \sin px \, \exp[i(\omega t - \beta z)]$$

$$\mathscr{H}_x = -\frac{\beta}{\omega\mu_0}\mathscr{E}_y$$

$$\mathscr{H}_z = \frac{ip}{\omega\mu_0}\cot px \, \mathscr{E}_y$$

for $|x| \leqslant a$ (2.66a)

$$\mathscr{E}_x = \mathscr{E}_z = 0$$

$$\mathscr{H}_y = 0$$

and

$$\mathscr{E}_y = \frac{x}{|x|}A_a \sin pa \, \exp[-\kappa(|x|-a)]$$

$$\times \exp[i(\omega t - \beta z)]$$

$$\mathscr{H}_x = -\frac{\beta}{\omega\mu_0}E_y$$

for $|x| \geqslant a$ (2.66b)

$$\mathscr{H}_z = -\frac{i\kappa}{\omega\mu_0}\frac{|x|}{x}E_y$$

$$\mathscr{E}_x = \mathscr{E}_z = 0$$

$$\mathscr{H}_y = 0$$

Symmetric TM modes

$$\mathcal{H}_y = B_s \cos px \, \exp[i(\omega t - \beta z)]$$

$$\mathcal{E}_x = -\frac{i\beta}{\omega \varepsilon_1} \mathcal{H}_y$$

$$\mathcal{E}_z = \frac{ip}{\omega \varepsilon_1} \tan px \, \mathcal{H}_y \qquad\qquad \text{for } |x| \leq a \qquad (2.67a)$$

$$\mathcal{H}_x = \mathcal{H}_z = 0$$

$$\mathcal{E}_y = 0$$

and

$$\mathcal{H}_y = B_s \cos pa \, \exp[-\kappa(|x|-a)]$$
$$\times \exp[i(\omega t - \beta z)]$$

$$\mathcal{E}_x = -\frac{i\beta}{\omega \varepsilon_2} \mathcal{H}_y$$

$$\qquad\qquad\qquad\qquad\qquad\qquad \text{for } |x| \geq a \qquad (2.67b)$$

$$\mathcal{E}_z = \frac{i\kappa}{\omega \varepsilon_2} \frac{|x|}{x} \mathcal{H}_y$$

$$\mathcal{H}_x = \mathcal{H}_z = 0$$

$$\mathcal{E}_y = 0$$

Antisymmetric TM modes

$$\mathcal{H}_y = B_a \sin px \, \exp[i(\omega t - \beta z)]$$

$$\mathcal{E}_x = -\frac{i\beta}{\omega \varepsilon_1} \mathcal{H}_y$$

$$\mathcal{E}_z = -\frac{ip}{\omega \varepsilon_1} \cot px \, \mathcal{H}_y \qquad\qquad \text{for } |x| \leq a \qquad (2.68a)$$

$$\mathcal{H}_x = \mathcal{H}_z = 0$$

$$\mathcal{E}_y = 0$$

and

$$\mathcal{H}_y = B_a \sin pa \frac{|x|}{x} \exp[-\kappa(|x|-a)]$$

$$\times \exp[i(\omega t - \beta z)]$$

$$\mathcal{E}_x = -\frac{i\beta}{\omega\varepsilon_2} \mathcal{H}_y$$

$$\mathcal{E}_z = \frac{i\kappa}{\omega\varepsilon_2} \frac{|x|}{x} \mathcal{H}_y \qquad\qquad \text{for } x|x \geqslant a \qquad (2.68b)$$

$$\mathcal{H}_x = \mathcal{H}_z = 0$$

$$\mathcal{E}_y = 0$$

The propagation constants, β_n, are solutions of the following transcendental equations

$$\tan pa = \frac{\kappa}{p} \qquad\qquad \text{symmetric TE modes} \qquad\qquad (2.69)$$

$$\tan pa = -\frac{p}{\kappa} \qquad\qquad \text{antisymmetric TE modes} \qquad\qquad (2.70)$$

$$\tan pa = \frac{\varepsilon_1}{\varepsilon_2}\frac{\kappa}{p} \qquad\qquad \text{symmetric TM modes} \qquad\qquad (2.71)$$

$$\tan pa = -\frac{\varepsilon_2}{\varepsilon_1}\frac{p}{\kappa} \qquad\qquad \text{antisymmetric TM modes} \qquad\qquad (2.72)$$

where

$$p^2 = \omega^2 \varepsilon_1 \mu_0 - \beta^2 \qquad\qquad (2.73a)$$

and

$$\kappa^2 = \beta^2 - \omega^2 \varepsilon_2 \mu_0 \qquad\qquad (2.73b)$$

The above equations can be rewritten in the form

$$\cot \alpha = \pm \frac{\alpha}{(g^2 - \alpha^2)^{1/2}} \qquad\qquad \text{symmetric TE modes} \qquad\qquad (2.74)$$

$$\tan \alpha = \pm \frac{\alpha}{(g^2 - \alpha^2)^{1/2}} \qquad\qquad \text{antisymmetric TE modes} \qquad\qquad (2.75)$$

$$\cot \alpha = \pm \frac{\varepsilon_2}{\varepsilon_1} \frac{\alpha}{(g^2 - \alpha^2)^{1/2}} \qquad\qquad \text{symmetric TM modes} \qquad\qquad (2.76)$$

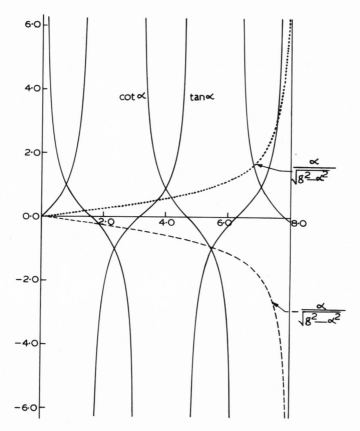

Fig. 2.2. Graphical method for solving Eqs. (2.74) and (2.75) for $g = 8$. The points of intersection of the solid curves with the dashed (and dotted) curves give the values of α.

and

$$\tan \alpha = \pm \frac{\varepsilon_2}{\varepsilon_1} \frac{\alpha}{(g^2 - \alpha^2)^{1/2}} \qquad \text{antisymmetric TM modes} \qquad (2.77)$$

where

$$\alpha = pa = (\omega^2 \varepsilon_1 \mu_0 - \beta^2)^{1/2} a \qquad (2.78a)$$

and

$$g^2 = \omega^2 \mu_0 (\varepsilon_1 - \varepsilon_2) a^2 \qquad (2.78b)$$

For given values of ω, ε_1, ε_2, and a, graphical solutions of Eqs. (2.74)–(2.77) are quite straightforward. For example, for $g = 8$, the LHS (left-hand side) and RHS of Eqs. (2.74) and (2.75) are plotted in Fig. 2.2. The roots are given

and

$$k_{1z} = k_1 \sin \theta_N = \beta_N \qquad (2.85)$$

Since

$$\alpha_N < g \qquad (2.86)$$

we will have

$$k_1 \cos \theta_N < g/a \qquad (2.87)$$

or

$$\cos \theta_N < \frac{g}{ak_1} = \left(1 - \frac{\varepsilon_2}{\varepsilon_1}\right)^{1/2} \qquad (2.88)$$

Alternatively

$$\sin \theta_N > (\varepsilon_2/\varepsilon_1)^{1/2} = n_2/n_1 \qquad (2.89)$$

where n_1 and n_2 are the refractive indices of regions I and II, respectively. Equation (2.89) implies that

$$\theta_N > \theta_c \qquad (2.90)$$

where $\theta_c [= \sin^{-1}(n_2/n_1)]$ is the critical angle. Thus the guided waves undergo total internal reflection at the interface.

It should be pointed out that until now we have been discussing modes which decrease exponentially in the region $|x| > a$. There exists a continuum of radiation modes which have oscillatory solution outside the slab; these correspond to $\beta^2 > \omega^2 \varepsilon_1 \mu_0$. The radiation modes along with the guided modes form the complete solution of the propagation problem. However, radiation modes are not relevant while considering the waveguiding properties of a system, but are important while considering radiation losses due to mode conversion (see Chapter 8).

2.4. Asymmetric Guides

In Appendix A we have discussed the experimental technique for fabrication of asymmetric metal–dielectric–dielectric layer structures. In this section we will attempt a modal analysis of asymmetric guides taking into account the effect of absorption; the treatment is based on the papers by Marcuse (1973a) and Kaminow *et al.* (1974).

As in the previous section we will assume the wave propagation to be along the z direction and the field patterns to be independent of y. The complex dielectric constants of the three regions can be written as

$$K_j = K_{rj} - iK_{ij} \qquad (2.91)$$

where the subscripts r and i refer to the fact that K_{rj} and K_{ij} are the real and imaginary parts of the complex dielectric constant K_j; and $j = 1, 2, 3$ refer to the three regions $-d < x < 0$, $0 < x < \infty$, and $-\infty < x < -d$, respectively. For waveguiding action we must have (see Fig. 2.4)

$$K_{r1} > K_{r2} \geqslant K_{r3} \qquad (2.92)$$

The complex refractive index $n_j = n_{rj} - in_{ij}$ is related to K_j by the following relation

$$n_j^2 = K_j$$

Thus

$$K_{rj} = n_{rj}^2 - n_{ij}^2 \qquad (2.93a)$$

and

$$K_{ij} = 2n_{rj}n_{ij} \qquad (2.93b)$$

The wave equation in each region can be solved in a manner similar to that in the previous section, except for the fact that the modes cannot, any longer, be classified as symmetric and antisymmetric.

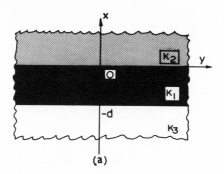

(a)

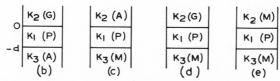

Fig. 2.4. (a) Cross section of a planar waveguide with $K_1 > K_2 > K_3$ (b) air/polymer/glass (APG) guide (c) air/polymer/metal (APM) guide (d) metal/polymer/glass (MPG) guide (e) metal/polymer/metal (MPM) guide (after Kaminow et al., 1974; reprinted by permission).

For TE modes once again the only nonvanishing field components would be H_z, H_x, and E_y; E_y would satisfy [see, e.g., Eq. (2.44)]

$$\frac{d^2E_y}{dx^2} + \left(\frac{\omega^2}{c^2}K_j - \beta^2\right)E_y = 0 \qquad j = 1, 2, 3 \qquad (2.94)$$

and H_z and H_x would be related to E_y through Eqs. (2.42) and (2.43). For guided propagation the fields should decay exponentially in the regions $0 < x < \infty$ and $-\infty < x < -d$; thus we obtain

$$E_y = A \, \exp[-\gamma x] \hspace{4cm} \text{for } 0 < x < \infty \qquad (2.95)$$

$$= A[\cos \kappa x - (\gamma/\kappa) \sin \kappa x] \hspace{2cm} \text{for } -d < x < 0 \qquad (2.96)$$

$$= A[\cos \kappa d + (\gamma/\kappa) \sin \kappa d] \exp[\theta(x+d)] \hspace{1cm} \text{for } -\infty < x < -d \qquad (2.97)$$

where use has been made of the condition that E_y is continuous at $x = 0$ and at $x = -d$; and that $H_z[=(i/\omega\mu_0)(\partial E_y/\partial x)]$ is continuous at $x = 0$. Further

$$\gamma^2 = \beta^2 - K_2 k_0^2 \qquad (2.98)$$

$$\theta^2 = \beta^2 - K_3 k_0^2 \qquad (2.99)$$

$$\kappa^2 = K_1 k_0^2 - \beta^2 \qquad (2.100)$$

where $k_0^2 = \omega^2/c^2$. Continuity of $H_z[=(i/\omega\mu_0)(\partial E_y/\partial x)]$ at $x = -d$ gives the transcendental equation

$$\tan \kappa d = \frac{\kappa(\gamma + \theta)}{\kappa^2 - \gamma\theta} \qquad (2.101)$$

the solution of which gives the propagation constants. In general, γ, β, θ, and κ will be complex:

$$\gamma = \gamma_r - i\gamma_i$$
$$\beta = \beta_r - i\beta_i$$
$$\theta = \theta_r - i\theta_i \qquad (2.102)$$
$$\kappa = \kappa_r - i\kappa_i$$

where once again the subscripts r and i refer to the real and imaginary parts, respectively. Obviously, for the fields to be bounded, β_i, γ_r, and θ_r should be positive. The power carried by the mode is related to the electric field through the relation [Eq. (2.56)]

$$P = \frac{1}{2}\frac{\beta_r}{\omega\mu_0}\int_{-\infty}^{\infty} |E_y|^2 \, dx \qquad (2.103)$$

Similarly, for TM modes the only nonvanishing field components would be H_y, E_z, and E_x; H_y would be given by

$$H_y = D \exp[-\gamma x] \qquad\qquad 0 < x < \infty \qquad (2.104)$$

$$= D\left(\cos \kappa x - \frac{K_1}{K_2}\frac{\gamma}{\kappa}\sin \kappa x\right) \qquad\qquad -d < x < 0 \qquad (2.105)$$

$$= D\left(\cos \kappa d + \frac{K_1}{K_2}\frac{\gamma}{\kappa}\sin \kappa d\right)\exp[\theta(x+d)] \qquad -\infty < x < -d \qquad (2.106)$$

where use has been made of the condition that H_y is continuous at $x = 0$ and at $x = -d$, and that $E_z[=(-i/\omega\varepsilon_0 K)(\partial H_y/\partial x)]$ is continuous at $x = 0$. Continuity of E_z at $x = -d$ gives

$$\tan \kappa d = \frac{K_1\kappa(K_3\gamma + K_2\theta)}{K_2 K_3 \kappa^2 - K_1^2\gamma\theta} \qquad (2.107)$$

which would determine the propagation constants for various modes. The power carried by the mode is given by

$$P = \tfrac{1}{2}\mathrm{Re}\left(\frac{\beta}{\omega\varepsilon_0}\int_{-\infty}^{\infty} K_j^{-1}|H_y|^2\,dx\right) \qquad (2.108)$$

Equations (2.101) and (2.107) can be put in the following convenient form

$$k_0 d = (K_1 - B)^{-1/2}\left\{\tan^{-1}\left[f\left(\frac{B-K_2}{K_1-B}\right)^{1/2}\right]\right.$$

$$\left. + \tan^{-1}\left[g\left(\frac{B-K_3}{K_1-B}\right)^{1/2}\right] + \nu\pi\right\} \qquad (2.109)$$

where

$$B = (\beta/k_0)^2$$

$$f = g = 1 \qquad\qquad \text{for TE modes}$$

$$f = K_1/K_2 \qquad\qquad\qquad\qquad\qquad\qquad\qquad (2.110)$$

$$\qquad\qquad\qquad\qquad \text{for TM modes}$$

$$g = K_1/K_3$$

and modes are labeled $\nu = 0, 1, 2, \ldots$ for solutions having decreasing values of β^2. Further, in Eq. (2.109), the real part of the arctan function is assumed to lie between $-(\pi/2)$ and $+(\pi/2)$. Kaminow et al. (1974) have developed a computer program to calculate the complex propagation constants when K_3 and/or K_2 are complex. The complex dielectric constants for air (A), polymer film (P), and the glass substrate (G) were assumed to be $1 - i0, 2.523 - i0$, and $2.290 - i0$ respectively. For the metal layer (M) the

Table 2.1. Comparison of Measured and Calculated Parameters[a]

Sample	$k_0 d$	Mode	$(\beta_r/k_0)_{meas}$	$(\beta_r/k_0)_{calc}$	α_{meas} (dB/cm)	$(\beta_i/k_0)_{meas}$ $(\times 10^{-7})$	$(\beta_i/k_0)_{calc}$ $(\times 10^{-7})$
APAg	59	TE_0	1.587	1.588	0.6	7	0.9
		TE_1	1.583	1.583	0.8	9	3.5
		TE_2	1.578	1.580	0.7	8	8
		TM_0	—	1.726	—	—	5×10^4
		TM_1	1.586	1.588	1.2	14	8
APAg	26	TE_0	1.578	1.583	1.0	12	11
		TE_1	1.569	1.569	~3	35	45
		TE_2	1.552	1.550	~10	120	100
		TM_0	—	1.718	—	—	5×10^4
		TM_1	1.577	1.582	~8	93	110
		TM_2	1.566	1.568	~15	174	400
APAg	10	TE_0	1.56	1.56	—	—	150
		TE_1	1.49	1.48	—	—	700
		TM_0	—	1.73	—	—	5×10^4
		TM_1	1.55	1.55	—	—	2×10^3
		TM_2	1.45	1.47	—	—	5×10^3
AgPG	59	TM_0	1.71	1.726	—	—	5×10^4
APAu	72	TE_0	1.588	1.588	0.5	6	2
		TE_1	1.585	1.585	1.5	17	8
		TE_2	1.582	1.582	2.1	24	18
		TM_0	—	1.825	—	—	3×10^5
		TM_1	1.586	1.587	1.6	19	13
		TM_2	1.581	1.582	8.2	95	50
APAl	59	TE_0	1.588	1.588	3.4	39	7
		TE_1	1.583	1.583	8.6	100	30
		TE_2	1.576	1.580	13.3	154	70
		TM_0	—	1.634	—	—	2×10^5
		TM_1	1.586	1.587	3.5	41	190
		TM_2^-	1.581	1.581	5.8	67	630

[a] After Kaminow et al. (1974).

dielectric constant was $-16.32 - i0.5414$ for silver, $-10.28 - i1.040$ for gold, and $-39.88 - i15.56$ for aluminum.* The dielectric constants correspond to $\lambda_0 = 6330$ Å. The actual metal thickness was around $0.1\ \mu m$; this was assumed to be infinite in the calculation of β (the error involved is negligible). The comparison of measured and calculated parameters is given in Table 2.1. The measurements were carried out at $\lambda_0 = 6330$ Å ($k_0 = 9.29 \times 10^4\ cm^{-1}$). The power attenuation coefficient (in dB/cm) is related to β_i (measured in cm^{-1}) through the following relation

$$\alpha = 8.686\ \beta_i$$

*Note that at optical frequencies the real part of the dielectric constant for a metal is negative.

Kaminow *et al.* (1974) have pointed out that since the waveguide propagation and measurements of phase velocity and attenuation are quite simple, one may use the inverse technique for determination of the complex dielectric constants.

CHAPTER 3

Planar Waveguides with Continuous Dielectric Constant Variation

In the last chapter, we discussed electromagnetic wave propagation in planar slab waveguides consisting of a slab of high refractive index material sandwiched between two layers. From the point of view of geometrical optics a light beam is guided by repeated total internal reflections at the interfaces. One of the disadvantages of such waveguides is the rapid increase in the transmission loss with increasing irregularities of the interfaces; this places rigid tolerance limits for low-loss waveguides. Recently, much interest has been aroused in waveguides in which the guidance is the result of continuous variation of refractive index. The fabrication of such waveguides is discussed in Appendix A. The waveguides act as a cylindrical lens with periodic focusing of the beam; most of the energy of the beam is concentrated around the central portion (where the refractive index is maximum) and hence the effect of irregularities (at the peripheries) on the beam is negligible.

In the first part of this chapter we discuss guided propagation in a medium characterized by a parabolic refractive index distribution, i.e.,

$$n = n_0 - n_2 x^2$$

In the second part we discuss the waveguiding action in a medium having an exponential variation of refractive index

$$n = n_1 + \Delta n \, \exp(-x/d) \qquad \text{for } x > 0$$

and

$$n = n_0 \qquad \text{for } x < 0$$

In the third part a continuous dielectric constant model (applicable to p–n junctions) is considered.

The discussion includes the application of these analyses to realistic situations.

3.1. *The Square-Law Medium*

In this part we will be discussing transmission in a medium characterized by the refractive index variation

$$n = n_0 - n_2 x^2 \qquad (3.1)$$

Such a medium is often termed a square-law medium. It has been shown by Miller (1965) that only in an aberrationless-lens waveguides or in a continuous medium with square-law refractive index variation will the shape of a beam injected off-axis (or with an angle to the medium's axis) remain invariant about a beam axis which oscillates about the axis of the medium. In a non-square-law medium the beam will, in general, spread. It should be noted that Eq. (3.1) cannot be valid for large values of x because it predicts negative values for the refractive index as $x \to \infty$. However, as long as the beam is confined to a region for which $n_2 x^2 \ll n_0$, use of Eq. (3.1) for all values of x will lead to negligible error. This is often termed as the infinite medium approximation. The effect of the presence of a boundary will be considered in Chapter 6. We should mention that the refractive index variation of the type described by Eq. (3.1) is rather difficult to obtain experimentally, except for some special waveguides of the type fabricated by Izawa and Nakagome (1972); however, theoretical studies on propagation in square-law media are essential to an understanding of guidance in more complicated cases. Further, for a slab waveguide characterized by the refractive index distribution of the form given by Eq. (3.1), one can solve the wave equation analytically, which is not possible in the general case. The analysis also enables us to analytically estimate the error involved in neglecting the term involving $\nabla(\ln K)$ in the wave equation.

3.1.1. Modal Analysis

Assuming the z and t dependence of the fields to be of the form

$$\exp[i(\omega t - \beta z)]$$

one obtains [see Eqs. (2.22b) and (2.24a) with $\gamma = 0$]

$$\frac{d^2 E_y}{dx^2} + \left[\frac{\omega^2}{c^2} K(x) - \beta^2\right] E_y = 0 \tag{3.2}$$

for TE modes* and

$$\frac{d^2 \phi}{dx^2} + \left[\frac{\omega^2}{c^2} K(x) - \frac{3}{4}\left(\frac{1}{K}\frac{dK}{dx}\right)^2 + \frac{1}{2K}\frac{d^2 K}{dx^2} - \beta^2\right] \phi(x) = 0 \tag{3.3}$$

for TM modes. Here

$$\phi(x) = [K(x)]^{1/2} E_x(x)$$

and for the present problem

$$K = n^2 \simeq K_0 - K_2 x^2$$
$$K_0 = n_0^2 \tag{3.4}$$
$$K_2 = 2 n_0 n_2$$

On substituting the expression for $K(x)$ from Eq. (3.4) in Eq. (3.2) we obtain

$$\frac{d^2 E_y}{dx^2} + \left[\left(\frac{\omega^2}{c^2} K_0 - \beta^2\right) - \frac{\omega^2}{c^2} K_2 x^2\right] E_y = 0$$

which can be rewritten in the form

$$\frac{d^2 E_y}{d\xi^2} + (\lambda - \xi^2) E_y(\xi) = 0 \tag{3.5}$$

where

$$\xi = x/\alpha$$
$$\alpha = (c^2/\omega^2 K_2)^{1/4} \tag{3.6a}$$

and

$$\lambda = [(\omega^2/c^2) K_2]^{-1/2} [(\omega^2/c^2) K_0 - \beta^2] \tag{3.6b}$$

Equation (3.5) is of the same form as the one-dimensional Schrödinger equation for the linear harmonic oscillator problem [see, for example, Schiff

---

*In writing Eq. (3.2) we have used the fact that for TE modes E_y is proportional to H_x [see Eq. (2.42)].

(1968)]. For the solution to be bounded for ξ tending to $\pm\infty$, we must have

$$\lambda = (2n+1) \qquad n = 0, 1, 2, \ldots \qquad (3.7)$$

The corresponding wave function would be Hermite–Gaussian functions:

$$E_y = N_n H_n(\xi) \exp(-\tfrac{1}{2}\xi^2) \qquad (3.8)$$

where $H_n(\xi)$ represents the Hermite polynomials and N_n is an arbitrary constant;* thus E_y may, in general, be expressed as

$$E_y = \sum_{n=0}^{\infty} A_n N_n H_n(\xi) \exp(-\tfrac{1}{2}\xi^2) \exp(-i\beta_n z)$$

The first few Hermite polynomials are of the form

$$H_0(\xi) = 1$$
$$H_1(\xi) = 2\xi$$
$$H_2(\xi) = 4\xi^2 - 2$$
$$H_3(\xi) = 8\xi^3 - 12\xi$$

The first and the second modal patterns are plotted in Fig. 3.1.

Using Eqs. (3.6) and (3.7) one obtains the following expression for the propagation constants for the TE modes:

$$\beta_n^2 = (\omega^2/c^2)K_0 - (2n+1)[(\omega^2/c^2)K_2]^{1/2} \qquad (3.9)$$

For TM modes, $\phi(x)$ satisfies the following equation:

$$\frac{d^2\phi}{dx^2} + \left\{ \left(\frac{\omega^2}{c^2}K_0 - \beta^2 - \frac{K_2}{K_0} \right) - \left[\frac{\omega^2}{c^2}K_2 + 4\left(\frac{K_2}{K_0}\right)^2 \right]x^2 \right\}\phi = 0 \qquad (3.10)$$

where terms of $O[(K_2/K_0)^3 x^4]$ have been neglected. Equation (3.10) can also be transformed to an equation of the type of Eq. (3.5) and hence the eigenfunctions would again be Hermite–Gaussian functions; thus

$$E_x = \frac{1}{[K(x)]^{1/2}}\phi(x) \sim \frac{1}{[K(\eta)]^{1/2}}H_n(\eta)\exp(-\tfrac{1}{2}\eta^2) \qquad (3.11)$$

*If we choose

$$N_n = \left[\frac{1/\alpha}{2^n n! \sqrt{\pi}} \right]^{1/2}$$

we can show that $\int_{-\infty}^{+\infty} |E_y(x)|^2 \, dx = 1$; this choice offers some mathematical advantages. It may be mentioned that the symbol H is used both for the magnetic field as well as for Hermite polynomials so one should be careful! However, the former will have subscripts x and y, whereas the latter will have n or m.

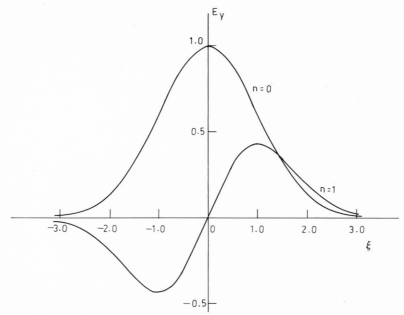

Fig. 3.1. The modal patterns corresponding to $n = 1$ and $n = 2$. The functions are simply the Hermite–Gauss functions.

where

$$\eta = x/\alpha' \tag{3.12a}$$

$$\alpha' = \left[\frac{\omega^2}{c^2}K_2 + 4\left(\frac{K_2}{K_0}\right)^2\right]^{-1/4}$$

$$= \left(\frac{\omega^2}{c^2}K_2\right)^{1/4}\left(1 + \frac{4K_2}{K_0^2}\frac{c^2}{\omega^2}\right)^{-1/4} \tag{3.12b}$$

Further, the propagation constants are given by

$$\beta_n^2 = (\omega^2/c^2)K_0 - (K_2/K_0) - (2n+1)[(\omega^2/c^2)K_2 + 4(K_2/K_0)^2]^{1/2} \tag{3.13}$$

If we had neglected the term depending on ∇K, then E_x would have satisfied an equation similar to E_y [Eq. (3.2)] and the corresponding propagation constants would have been given by

$$\beta_n^2 = (\omega^2/c^2)K_0 - (2n+1)[(\omega^2/c^2)K_2]^{1/2} \tag{3.13a}$$

Comparison of Eq. (3.13) with (3.13a) shows that the effect of the term depending on $\nabla(\ln K)$ is negligible if $(K_2/K_0)(c^2/\omega^2) \ll 1$; i.e., if the characteristic distance for the variation of the refractive index is large compared to the wavelength. This is indeed true for practical systems at optical frequencies. Equation (3.13) can be rewritten in the form

$$\beta_n = \frac{\omega}{c} K_0^{1/2} \left[1 - \frac{c^2}{\omega^2} \frac{1}{K_0} \left(\frac{K_2}{K_0} \right) \right.$$

$$\left. - (2n+1) \frac{c}{\omega} \frac{K_2^{1/2}}{K_0} \left(1 + \frac{4K_2}{K_0^2} \frac{c^2}{\omega^2} \right)^{1/2} \right]^{1/2}$$

$$\approx \frac{\omega}{c} K_0^{1/2} \left[1 - \frac{c^2}{\omega^2} \frac{1}{K_0} \left(\frac{K_2}{K_0} \right) \right.$$

$$\left. - (2n+1) \frac{c}{\omega} \frac{K_2^{1/2}}{K_0} - 2(2n+1) \left(\frac{c}{\omega} \right)^3 \frac{K_2^{3/2}}{K_0^3} + \cdots \right]^{1/2}$$

$$\tag{3.14}$$

$$\approx \frac{\omega}{c} K_0^{1/2} \left\{ 1 - (n + \tfrac{1}{2}) \frac{c}{\omega} \frac{K_2^{1/2}}{K_0} \right.$$

$$\left. - \frac{c^2}{\omega^2} \left[\frac{1}{2K_0} \left(\frac{K_2}{K_0} \right) - \tfrac{1}{8}(2n+1)^2 \frac{K_2}{K_0} \right] + \cdots \right\}$$

$$\approx \frac{\omega}{c} K_0^{1/2} - (n + \tfrac{1}{2}) \frac{K_2}{K_0} - O\left(\frac{c}{\omega} \right)$$

[The O in expression (3.14) indicates "of order."] Now the group velocity is $1/(d\beta/d\omega)$ and it can immediately be seen that if terms of $O(c/\omega)$ are neglected then $d\beta_n/d\omega$ is independent of n; i.e., all the modes have the same group velocity. This is one of the most important characteristics of a square-law medium (see also Chapter 7) and leads to negligible distortion of a propagating pulse.

In what follows we will assume $(K_2/K_0)(c/\omega)^2$ to be very small compared to unity and will, therefore, neglect the term depending on $\nabla(\ln K)$. Thus, it will be sufficient to consider the solution of the wave equation

$$\nabla^2 \psi + \frac{\omega^2}{c^2} K \psi = 0 \tag{3.15}$$

where ψ may be assumed to represent any one of the field components.

3.1.2.   The Kernel

The mathematical analysis of propagation is much simplified by the introduction of a function known as a kernel. As shown later in this section, if $\phi(x, z = 0)$ represents the field pattern at $z = 0$, then the field distribution at an arbitrary value of z would be given by

$$\phi(x, z) = \int K(x, x', z)\phi(x', z = 0)dx'$$

where the function K is the kernel. Thus the evaluation of the kernel will enable us to determine the field pattern at an arbitrary value of z in terms of the field distribution at $z = 0$. In order to determine the kernel we note that the field patterns corresponding to the various modes are given by [see Eqs. (3.8) and (3.9)]

$$\psi_n = N_n H_n(\xi) \exp(-\tfrac{1}{2}\xi^2) \tag{3.16}$$

where

$$\beta_n = \left[\frac{\omega^2}{c^2}K_0 - (2n+1)\left(\frac{\omega^2}{c^2}K_2\right)^{1/2}\right]^{1/2} \tag{3.17}$$

Thus, the general solution of Eq. (3.15) would be given by

$$\phi(x, z) = \sum_n A_n \psi_n(x) \exp(-i\beta_n z) \tag{3.18}$$

Let $\phi(x, z = 0)$ denote the field distribution at $z = 0$, then

$$\phi(x, z = 0) = \sum_n A_n \psi_n(x)$$

By using the orthonormality condition*

$$\int_{-\infty}^{+\infty} \psi_m^*(x)\psi_n(x) = \delta_{mn}$$

we obtain

$$A_n = \int_{-\infty}^{+\infty} \psi_n^*(x)\phi(x, z = 0)\, dx$$

Thus

$$\phi(x, z) = \sum_n \left[\int_{-\infty}^{+\infty} \psi_n^*(x')\phi(x, z = 0)\, dx'\right]\psi_n(x) \exp(-i\beta_n z)$$

$$= \int_{-\infty}^{+\infty} K(x, x', z)\phi(x', z = 0)\, dx' \tag{3.19}$$

where

$$K(x, x', z) = \sum_n \psi_n^*(x')\psi_n(x) \exp(-i\beta_n z) \tag{3.20}$$

*$\delta_{mn} = 0$ if $n \neq m$; $\delta_{mn} = 1$ if $n = m$.

represents the kernel. Equation (3.19) is of the same form as those encountered in diffraction theory [see, for example, Born and Wolf (1970)].

The summation appearing in Eq. (3.20) can be carried out provided we neglect terms of $O[(c/\omega)(K_2^{1/2}/K_0)]$ in the binomial expansion of the RHS of Eq. (3.17):

$$
\begin{aligned}
\beta_n &\simeq \frac{\omega}{c} K_0^{1/2} \left[1 - (n + \tfrac{1}{2}) \frac{c}{\omega} \frac{K_2^{1/2}}{K_0} \right] \\
&\simeq \frac{\omega}{c} K_0^{1/2} - (n + \tfrac{1}{2}) \left(\frac{K_2}{K_0} \right)^{1/2}
\end{aligned}
\tag{3.21}
$$

Substituting the expressions for ψ_n from Eq. (3.16) and for β_n from Eq. (3.21) in Eq. (3.20) we obtain

$$
\begin{aligned}
K(x, x', z) &= \sum_n \left(\frac{1/\alpha}{2^n n! \sqrt{\pi}} \right) H_n(\xi) H_n(\xi') \exp[-\tfrac{1}{2}(\xi^2 + \xi'^2)] \\
&\quad \times \exp\left\{ -i \left[\frac{\omega}{c} K_0^{1/2} - (n + \tfrac{1}{2}) \left(\frac{K_2}{K_0} \right)^{1/2} \right] z \right\} \\
&= \frac{1}{\alpha \sqrt{\pi}} \exp\left\{ -\tfrac{1}{2}(\xi^2 + \xi'^2) - i \left[\frac{\omega}{c} K_0^{1/2} - \tfrac{1}{2} \left(\frac{K_2}{K_0} \right)^{1/2} \right] z \right\} \\
&\quad \times \sum_{n=0,1,2,\ldots} \frac{\zeta^n}{2^n n!} H_n(\xi') H_n(\xi)
\end{aligned}
\tag{3.22}
$$

where

$$
\zeta = \exp\left[+i \left(\frac{K_2}{K_0} \right)^{1/2} z \right]
\tag{3.23}
$$

Thus

$$
K(x, x', z) = \frac{1}{\alpha \sqrt{\pi} (1 - \zeta^2)^{1/2}} \exp\left\{ -\tfrac{1}{2}(\xi^2 + \xi'^2) - i\gamma z \right. \\
\left. + \left[\frac{2\xi\xi'\zeta - (\xi^2 + \xi'^2)\zeta^2}{1 - \zeta^2} \right] \right\}
\tag{3.24}
$$

where

$$
\gamma = (\omega/c) K_0^{1/2} - \tfrac{1}{2}(K_2/K_0)^{1/2}
$$

and use has been made of Mehler's formula [see, for example, the Bateman Manuscript Project (1953)] for carrying out the summation which appears on the RHS of Eq. (3.22). Substituting for K from Eq. (3.24) in Eq. (3.19) one obtains

$$\phi(x, z) = \frac{\exp(-i\gamma z)}{\sqrt{\pi}(1-\zeta^2)^{1/2}} \int_{-\infty}^{+\infty} \phi(\xi', z = 0)$$

$$\times \exp\left[\frac{2\xi\xi'\zeta}{1-\zeta^2} - \frac{(\xi^2+\xi'^2)(1+\zeta^2)}{2(1-\zeta^2)}\right] d\xi' \tag{3.25}$$

Thus, if the field distribution at the entrance aperture of the guide $(z = 0)$ is known, i.e., if $\phi(\xi, z = 0)$ is known, then the field distribution at an arbitrary point z can be easily obtained by carrying out the integration in Eq. (3.25). It should be noted that we have made only two approximations in the derivation of Eq. (3.25): (i) neglect of the term involving $\nabla(\ln K)$ and (ii) the use of the binomial expansion of β_n.

We consider a specific example in which the incident field distribution is Gaussian, i.e.,

$$\phi(\xi, z = 0) = \phi_0 \exp[-\tfrac{1}{2}(\xi^2/\sigma^2)] \tag{3.26}$$

Thus

$$\phi(x, z) = A \int_{-\infty}^{+\infty} \exp\left[\frac{2\xi\xi'\zeta}{1-\zeta^2} - \frac{\xi'^2}{2}\left(\frac{1+\zeta^2}{1-\zeta^2} + \frac{1}{\sigma^2}\right)\right] d\xi' \tag{3.27}$$

where

$$A = \frac{\phi_0 \exp(-i\gamma z)}{\sqrt{\pi}(1-\zeta^2)^{1/2}} \exp\left[-\frac{\xi^2}{2}\left(\frac{1+\zeta^2}{1-\zeta^2}\right)\right]$$

Equation (3.27) can be put in the form

$$\phi(x, z) = A \exp(Q^2) \int_{-\infty}^{+\infty} \exp[-(P\xi' - Q)^2] d\xi'$$

$$= \frac{A}{P}\sqrt{\pi} \exp(Q^2) \tag{3.28}$$

where

$$P^2 = \frac{\sigma^2(1+\zeta^2) + (1-\zeta^2)}{2(1-\zeta^2)\sigma^2}$$

and

$$Q^2 = \frac{2\xi^2\zeta^2\sigma^2}{(1-\zeta^2)[(1+\zeta^2)\sigma^2 + (1-\zeta^2)]}$$

Thus

$$\phi(x, z) = \frac{\phi_0 \sqrt{2}\sigma \exp(-i\gamma z)}{[(1+\zeta^2)\sigma^2 + (1-\zeta^2)]^{1/2}}$$

$$\times \exp\left\{-\frac{\xi^2}{2}\left[\frac{(1+\zeta^2)+\sigma^2(1-\zeta^2)}{(1-\zeta^2)+\sigma^2(1+\zeta^2)}\right]\right\} \tag{3.29}$$

Substituting $\zeta = \exp(i\delta z)$ [see Eq. (3.23)], where $\delta = (K_2/K_0)^{1/2}$, in Eq. (3.29), we get

$$\phi(x, z) = \frac{2^{1/4}\phi_0\sigma \exp[+i\chi(x, z)]}{[\sigma^4 + 1 + (\sigma^4 - 1) \cos 2\delta z]^{1/4}}$$

$$\times \exp\left\{-\frac{\xi^2 \sigma^2}{\sigma^4 + 1 + (\sigma^4 - 1) \cos 2\delta z}\right\} \tag{3.30}$$

where

$$\chi(x, z) = -\frac{\omega}{c}K_0^{1/2}z + \frac{\xi^2}{2}\frac{(\sigma^4 - 1) \sin 2\delta z}{[(\sigma^4 + 1) + (\sigma^4 - 1) \cos 2\delta z]} + \frac{1}{2}\tan^{-1}\left(\frac{\tan \delta z}{\sigma^2}\right)$$

or,

$$|\phi(x, z)|^2 = |\phi_0|^2 \frac{w_0}{w(z)} \exp\left[-\frac{x^2}{w^2(z)}\right] \tag{3.30a}$$

where

$$w^2(z) = (w_0^2/2\sigma^4)[(\sigma^4 + 1) + (\sigma^4 - 1) \cos 2\delta z] \tag{3.31}$$

and $w_0 = \alpha\sigma$, the incident beamwidth (at $z = 0$). Thus the intensity distribution remains Gaussian and the variation of the beamwidth of the beam with z is given by Eq. (3.31); see Fig. 3.2. It can be seen from Eq. (3.31) that at planes given by $z = n\pi/\delta$; $n = 1, 2, 3, \ldots$ the beamwidth equals the incident beamwidth. At planes given by $z = (n + \frac{1}{2})\pi/\delta$; $n = 0, 1, 2, \ldots$ the beamwidth is

$$w^2 = w_0^2/\sigma^4$$

Thus the beamwidth at these planes is greater than the incident beamwidth if $\sigma < 1$, i.e., if the incident beamwidth w_0 is less than α, which is the width of the fundamental mode. Similarly the beamwidth at these planes is less than that of the incident beam if $\sigma > 1$, i.e., if the incident beamwidth is greater than α.

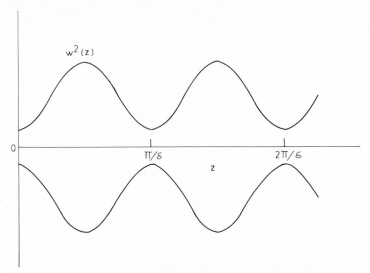

Fig. 3.2. The upper curve shows the variation of the square of the beam-width, $w^2(z)$, with z. The beamwidth repeats itself after a distance of π/δ. The distance between the two curves is a measure of the spot size of the beam.

A similar analysis can be carried out if the medium exhibits absorption. In such a case K_0 is complex and K_0 and ζ are of the form

$$K_0 = K_{0r} - iK_{0i}$$

$$\zeta = \exp(+i\delta z)\exp(-\kappa z)$$

where δ and κ are real. Further

$$\phi(x, z) = \frac{\phi_0\sqrt{2}\sigma \exp\left\{-\dfrac{\omega}{c}[K_{0i}/2(K_{0r})^{1/2}]z - \dfrac{1}{4}K_{0i}\left(\dfrac{K_2}{K_{0r}^3}\right)z\right\}\exp[-i\chi(x, z)]}{[\Lambda(z)]^{1/4}}$$

$$\times \exp\left\{-\frac{\xi^2[(\sigma^2+1)^2 - C^2(\sigma^2-1)^2]}{2\Lambda(z)}\right\} \qquad (3.32)$$

where

$$C = \exp(-2\kappa z)$$

$$\Lambda(z) = (\sigma^2+1)^2 + C^2(\sigma^2-1)^2 + 2C(\sigma^4-1)\cos 2\delta z$$

$$\chi(x, z) = \left(\frac{\omega}{c}(K_{0r})^{1/2} - \tfrac{1}{2}\delta\right)z - \frac{\xi^2 C(\sigma^4-1)\sin 2\delta z}{\Lambda(z)}$$

$$+ \tfrac{1}{2}\tan^{-1}\left[\frac{C(\sigma^2-1)\sin 2\delta z}{\sigma^2+1+C(\sigma^2-1)\cos 2\delta z}\right]$$

and we have assumed that $K_{0i} \ll K_{0r}$. The variation of the beamwidth with z is therefore given by

$$w^2(z) = \frac{w_0^2}{\sigma^2}\left[\frac{(\sigma^2+1)^2 + C^2(\sigma^2-1)^2 + 2C(\sigma^4-1)\cos 2\delta z}{(\sigma^2+1)^2 - C^2(\sigma^2-1)^2}\right] \qquad (3.33)$$

This equation reduces to Eq. (3.31) in the absence of absorption, namely as $C \to 1$. It is evident from Eq. (3.33) that in the limit $z \to \infty$, $C \to 0$ and the beamwidth tends to a value equal to the fundamental beamwidth of the medium.

3.1.3.   The Parabolic Equation Approach

When K_0 and K_2 are dependent on z, the modal analysis approach (for finding the intensity distribution as the beam propagates through the medium) becomes very cumbersome. An alternative method has recently been suggested (Sodha *et al.*, 1972b) which gives an approximate solution of the wave equation; this solution is in agreement with Eq. (3.30) when K_0 and K_2 are independent of z.

The mathematical technique is very similar to the one used for SELFOC fibers which will be discussed in detail in Chapter 5. We only give the results here[*]:

(i) In the geometrical optics approximation, the path of a ray in the meridional plane is represented by the equation

$$x = x_0 f(z) \qquad (3.34)$$

where $f(z)$ satisfies the following differential equation

$$\frac{d^2f}{dz^2} = -\frac{K_2}{K_0}f \qquad (3.35)$$

The boundary conditions are

$$f(z)|_{z=0} = 1$$

and $(1/f)(df/dz)|_{z=0}$ would represent the inclination of the ray at $z = 0$. For a ray that is incident normally, df/dz would vanish at the entrance aperture of the guide which is assumed to be at $z = 0$. The condition of $f = 1$ at $z = 0$ implies that x_0 represents the x-coordinate of the ray at $z = 0$. The solution of Eq. (3.35) can be found for a variety of z-dependence of K_2/K_0. In

---

[*]The intermediate steps can be found in the paper by Sodha *et al.*, 1972b.

particular, if K_2/K_0 is independent of z then for an incident ray parallel to the axis

$$f = \cos[(K_2/K_0)^{1/2}z]$$ (3.36)

Further, for $K_2(z)$ varying as $K_{20}(1+\beta z)$, one can find a closed form solution of Eq. (3.35) in terms of Bessel functions.

(ii) When the geometrical optics approximation is not valid then the intensity distribution of a beam, having a Gaussian profile at $z = 0$, remains Gaussian as the beam propagates in the medium and is given by

$$I = \frac{I_0}{f(z)} \exp\left(-\frac{x^2}{x_0^2 f^2}\right)$$ (3.37)

where $f(z)$ satisfies the following differential equation

$$\frac{1}{f}\frac{d^2f}{dz^2} + \frac{1}{f}\frac{df}{dz}\left(\frac{1}{k}\frac{dk}{dz}\right) = -\frac{K_2(z)}{K_0(z)} + \frac{1}{k^2 x_0^4 f^4}$$ (3.38)

and $k(z) = K_0(z)(\omega^2/c^2)$. The boundary conditions are

$$f|_{z=0} = 1$$ (3.39a)

and

$$\frac{1}{f}\frac{df}{dz}\bigg|_{z=0} = \frac{1}{R}$$ (3.39b)

where R is the radius of curvature of the incident cylindrical wave; the assumption $f = 1$ at $z = 0$ is without loss of generality and only implies that the width of the beam at $z = 0$ is x_0. When K_0 and K_2 are independent of z then solution of Eq. (3.38) can easily be found and is given by

$$f^2(z) = \frac{1}{2}\left\{(1+C) + (1-C)\cos 2\left[\left(\frac{K_2}{K_0}\right)^{1/2}z\right]\right\}$$ (3.40)

where $C = (K_0/K_2)(1/k^2 x_0^4)$ and we have assumed R to be ∞. Thus, a beam having Gaussian intensity distribution (which is symmetrical about the z-axis) remains Gaussian which is consistent with the conclusions of Sec. 3.4. Further, when $C = 1$, $f^2(z) = 1$ for all values of z and the beam propagates as the fundamental mode. Further we see that if we substitute for f^2 from Eq. (3.40) in Eq. (3.37) the resulting expression for the intensity is identical to that obtained from Eq. (3.33). This agreement provides confidence in the use of this method; hence it can be applied to the case when K_2 and K_0 are slowly varying functions of z.

3.2. Exponential Variation of the Refractive Index

Kaminow and Carruthers (1973) have fabricated waveguides (see Appendix A) in which the refractive index variation is of the form (see Fig. 3.3):

$$n(x) = n_0 \qquad\qquad \text{for } x < 0$$

$$= n_1 + \Delta n \, \exp\left(-\frac{x}{d}\right) \qquad \text{for } x > 0 \qquad (3.41)$$

where n_0 characterizes the homogeneous medium in the region $x < 0$; n_1 is the refractive index of LiNbO$_3$ (or LiTaO$_3$) and Δn represents the index change on the interface. Such a refractive index variation has also been obtained by diffusing Se into CdS crystals (Taylor *et al.*, 1972; Woodbury and Hall, 1967; Conwell, 1973).

This section presents a modal analysis of this type of waveguide. For TE modes propagating in the z-direction [following Conwell (1973)] we assume a solution of the form

$$\mathscr{E}_y = E_y(x) \, \exp[i(\omega t - \beta z)] \qquad (3.42)$$

where $E_y(x)$ satisfies the equation [see Eq. (3.2)]:

$$\frac{d^2 E_y}{dx^2} + \left[K(x)\frac{\omega^2}{c^2} - \beta^2 \right] E_y(x) = 0 \qquad (3.43)$$

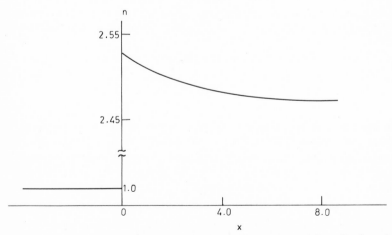

Fig. 3.3. Typical variation of refractive index for an exponential profile ($n_0 = 1$, $n_1 = 2.47$, $\Delta n/n_1 = 0.047$, and $d = 2.5\ \mu$).

For the present case the dielectric constant variation is given by

$$K(x) = K_0 = n_0^2 \qquad \text{for } x < 0$$
$$= K_1 + \Delta K \exp(-x/d) \qquad \text{for } x > 0 \qquad (3.44)$$

where $K_1 = n_1^2$ and $\Delta K = 2n_1 \Delta n$ and we have assumed $\Delta n \ll n_1$. In order to solve Eq. (3.43) in the region $x > 0$, we make the substitution

$$\xi = 4\pi (\Delta K)^{1/2} (d/\lambda) \exp(-x/2d) \qquad (3.45)$$

thus

$$\frac{dEy}{dx} = \frac{dEy}{d\xi} \cdot \frac{d\xi}{dx} = -\frac{dEy}{d\xi} \cdot \frac{2\pi}{\lambda} (\Delta K)^{1/2} \exp(-x/2d)$$

and

$$\frac{d^2 E_y}{dx^2} = \frac{1}{4d^2} \xi^2 \frac{d^2 E_y}{d\xi^2} + \frac{1}{4d^2} \xi \frac{dE_y}{d\xi}$$

On substituting the above expression for $(d^2 E_y/dx^2)$ in Eq. (3.43) we obtain

$$\xi^2 \frac{d^2 E_y}{d\xi^2} + \xi \frac{dE_y}{d\xi} + (\xi^2 - \alpha^2) E_y(\xi) = 0 \qquad (3.46)$$

where

$$\alpha^2 = 4d^2 \left(\beta^2 - \frac{K_1 \omega^2}{c^2} \right) \qquad (3.47)$$

The solutions of Eq. (3.46) are $J_\alpha(\xi)$ and $J_{-\alpha}(\xi)$; the latter diverges at $\xi = 0$ (which corresponds to $x = \infty$) and can hence be neglected.* Thus the modes of the system are given by

$$E_y(x) = A J_\alpha(\xi)$$
$$= A J_\alpha \left[4\pi (\Delta K)^{1/2} \frac{d}{\lambda} \exp(-x/2d) \right] \qquad (3.48)$$

The allowed values of the propagation constants β can be determined by applying the appropriate boundary conditions at $x = 0$. For $x < 0$, $K = K_0$ and one obtains the usual exponential solution

$$E_y(x) = B \exp(-p|x|) \qquad (3.49)$$

---

*It can easily be seen that α^2 cannot be negative; if we set $\gamma^2 = -\alpha^2$ then the solutions of Eq. (3.46) are $K_\gamma(\xi)$ and $I_\gamma(\xi)$; the former diverging at $\xi = 0$. If we use the solution $I_\gamma(\xi)$ then the derivative cannot be matched at $x = 0$ because $I_\gamma(\xi)$ is a monotonically increasing function of ξ.

where $p = (\beta^2 - K_0 \omega^2/c^2)^{1/2}$ and B is an arbitrary constant. The condition that E_y be continuous at $x = 0$ leads to a linear relation between A and B. Remembering that H_z is proportional to $\partial E_y/\partial x$ and requiring that H_z also be continuous at $x = 0$ leads to a second linear relation between A and B. These two linear relationships are given by

$$AJ_\alpha[4\pi(\Delta K)^{1/2}(d/\lambda)] = B$$

$$-A(\pi/\lambda)(\Delta K)^{1/2}[J_{\alpha-1}(\xi_0) - J_{\alpha+1}(\xi_0)] = Bp$$

where $\xi_0 = (4\pi/\lambda)(\Delta K)^{1/2}d$, and we have used the relation

$$\frac{d}{dx}\left\{J_\alpha\left[4\pi(\Delta K)^{1/2}\frac{d}{\lambda}\exp(-x/2d)\right]\right\}$$

$$= -\frac{4\pi(\Delta K)^{1/2}}{2\lambda}\exp(-x/2d)\frac{d}{d\xi}J_\alpha(\xi)$$

$$= -\frac{\pi}{\lambda}(\Delta K)^{1/2}\exp(-x/2d)[J_{\alpha-1}(\xi) - J_{\alpha+1}(\xi)]$$

Elimination of A and B gives us

$$\frac{J_{\alpha-1}(\xi_0) - J_{\alpha+1}(\xi_0)}{J_\alpha(\xi_0)} = -\frac{p\lambda}{\pi(\Delta K)^{1/2}} \tag{3.50}$$

3.2.1. TE Modes

The above equation specifies the possible β values for the TE modes. For guided modes $\beta^2 < (K_1 + \Delta K)\omega^2/c^2$; this follows from the fact that if $\beta^2 > (K_1 + \Delta K)\omega^2/c^2$, the field will be oscillatory when $x \to \infty$. Thus for the field to vanish for large values of x, β^2 must be less than $(K_1 + \Delta K)\omega^2/c^2$. Combining this condition with the lower limit on β obtained from the condition that α be real we obtain

$$(K_1 + \Delta K)^{1/2} > (c/\omega)\beta > K_1^{1/2} \tag{3.51}$$

This relation also ensures that p is real since $K_1 > K_0$. As it is obvious from the above equation, for $\Delta K \ll K_1$ (which is normally the case), that the range of β values is quite small. Since

$$p^2 = \beta^2 - (K_0\omega^2/c^2) \tag{3.52}$$

we obtain, using Eq. (3.51),

$$(K_1 - K_0 + \Delta K)^{1/2} > p\lambda/2\pi > (K_1 - K_0)^{1/2} \tag{3.53}$$

From Eq. (3.51) we also obtain

$$4\pi(\Delta K)^{1/2}(d/\lambda) > \alpha > 0 \qquad (3.54)$$

For the waveguides fabricated by Taylor *et al.* (1972), $K_1 = 6.1$, $K_0 = 1.0$, and the RHS of Eq. (3.50) varies only slightly over the range of α given by Eq. (3.54); further, for small values of ΔK (which is indeed the case), the RHS of Eq. (3.50) is quite large and therefore the characteristic values $\beta^{(i)}$ will correspond to those values of* $\alpha^{(i)}$, for which $J_\alpha[4\pi(\Delta K)^{1/2}(d/\lambda)] \to 0$. The number of TE modes the guide will support is then equal to the number of α values in the range given by Eq. (3.54), for which

$$J_\alpha[4\pi(\Delta K)^{1/2}(d/\lambda)] = 0 \qquad (3.55)$$

Conwell (1973) has calculated the number of modes for a waveguide, characterized by the following parameters

$$\Delta K = 0.3$$
$$\lambda = 0.633 \ \mu m \qquad (3.56)$$
$$d = 2.5 \ \mu m$$

The above parameters correspond to the system fabricated by Taylor *et al.* (1972). Thus

$$\xi_0 = 4\pi(\Delta K)^{1/2}(d/\lambda) \approx 27.0$$

and therefore α lies between 0 and 27.0. In this range of α, J_α (27.0) = 0, for $\alpha \approx 21.5$, 17.4, 14.1, 11.2, 8.5, 6.1, 3.8, and 1.7. Thus the guide will support eight TE modes. Assuming that the refractive index varies linearly with distance, Taylor *et al.* (1972) estimated that there will be only three TE modes. This is due to the fact that the linear approximation (assumed by Taylor *et al.*, 1972) underestimates the number of modes, since it arbitrarily cuts short the range of spatial variation of K.

It is easily seen that the solution expressed by Eq. (3.48) gives the correct asymptotic behavior as $x \to \infty$; in this limit the argument of the Bessel function (ξ) tends to zero and we know that

$$\lim_{\xi \to 0} J_\alpha(\xi) \to \xi^\alpha \qquad (3.57)$$

Thus for $x \to \infty$, $E_y \sim \exp(-x\alpha/2d)$, which is indeed the expected behavior.

For the waveguide characterized by Eq. (3.56) the modes may be numbered 0–7, in descending order of $\alpha^{(i)}$ and $\beta^{(i)}$. Figure 3.4 shows Conwell's calculations of the field patterns for the modes $m = 0$ and $m = 4$. It can be seen that the fundamental mode ($m = 0$) has no nodes† and the mode $m = 4$ has 4 nodes.

*α and β are related through Eq. (3.47).

†At nodes $E_y = 0$.

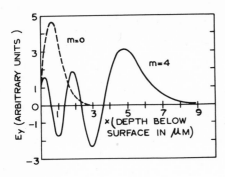

Fig. 3.4. Plots of $m = 0$ and $m = 4$ (unnormalized) modes calculated for $\lambda = 6328 \text{ Å}$ corresponding to a guide with an index of refraction varying according to $n = 2.47$ $[1 + 0.024 \exp(-x/2.5)]$ where x is in μm ($x > 0$) (after Conwell, 1973; reprinted by permission).

3.2.2.   TM Modes

For the TM modes, if we assume $\mathscr{H}_y$ to be of the same form as $\mathscr{E}_y$ given by Eq. (3.42), then

$$\frac{\partial}{\partial x}\left(\frac{1}{K(x)}\frac{\partial H_y}{\partial x}\right) + \left(\frac{\omega^2}{c^2} - \frac{\beta^2}{K(x)}\right)H_y = 0 \qquad (3.58)$$

Using the substitution

$$H_y = u(x)[K(x)]^{1/2}$$

and neglecting terms proportional to d^2K/dx^2 and $(dK/dx)^2$ one obtains a differential equation for $u(x)$ which is identical to Eq. (3.43). Thus the solution for H_y in the region $x > 0$ would be given by

$$H_y(x) \sim (K)^{1/2}J_\alpha[4\pi(\Delta K)^{1/2}(d/\lambda)\exp(-x/2d)] \qquad (3.59)$$

Using the continuity conditions and retaining terms of $O(\Delta K)$ one obtains

$$\frac{1}{J_\alpha(\xi_0)}\left[J'_\alpha(\xi_0) + \frac{\lambda(\Delta K)^{1/2}}{4\pi K_1 d}J_\alpha(\xi_0)\right]$$

$$= -\frac{K_1}{K_0}\frac{p\lambda}{2\pi(\Delta K)^{1/2}}\left(1 + \frac{\Delta K}{K_1}\right) \qquad (3.60)$$

where as before $\xi_0 = 4\pi(\Delta K)^{1/2}(d/\lambda)$. Conwell (1973) has shown that in spite of the difference in the structure of the transcendental equation [Eqs. (3.50) and (3.60)] the possible number of TM modes will almost invariably be the same as the possible number of TE modes, and the form of $H_y(x)$ will be quite similar to $E_y(x)$ since for cases of practical significance, the factor $[K(x)]^{1/2}$ varies very little with x.

3.3. A Continuous Dielectric Constant Model for p–n Junctions*

In Appendix A the nature of variation of the dielectric constant in a p–n junction has been indicated. Owing to the growth habit of crystals like GaP, it is experimentally convenient to fabricate diodes with the junction electric field,† E_{junct}, in the [111] direction. The advantages of this orientation have been discussed by Nelson and McKenna (1967). They have also suggested a continuously varying dielectric constant model which includes both the case where E_{junct} is in the [111] direction with the mode propagating in the perpendicular direction, and the case where E_{junct} is in the [100] direction with the mode propagating in the [011] direction. The x axis is taken along E_{junct} and the direction is assumed to be along the z axis. The dielectric constant tensor is diagonal in this coordinate system, and the diagonal elements which are functions of x only are assumed to be of the form

$$K_m(x) = K_{m0} - (K_{m0} - v_m)\left(\tanh\frac{x - \mu_m a}{a} + \tanh \mu_m\right)^2 \cosh^2 \mu_m \quad (3.61)$$

where $m = x, y, z$. The reason for choosing the above form is that it makes the analysis mathematically amenable and physically significant. The earlier analyses assumed a step discontinuity of dielectric constant across a plane which is physically unrealistic; the present model overcomes this deficiency.

For a crystal like GaP, with E_{junct} in the [111] direction, K_{m0} are given by (Nelson and McKenna, 1967):

$$K_{x0} = \frac{n^2}{1 - 2\delta} \approx n^2(1 + 2\delta) \quad (3.62a)$$

$$K_{z0} = K_{y0} = \frac{n^2}{1 + \delta} \approx n^2(1 - \delta) \quad (3.62b)$$

On the other hand, if E_{junct} is in the [100] direction, K_{m0} would be given by

$$K_{x0} = n^2 \quad (3.63a)$$

$$K_{y0} = \frac{n^2}{1 + \sqrt{3}\delta} \approx n^2(1 - \sqrt{3}\delta) \quad (3.63b)$$

---

*This section is based on a paper by Nelson and McKenna (1967).

†The current density J is given by $J = -|e|D_e \nabla n_e + |e|D_h \nabla n_h + \sigma E$, where the suffixes e and h refer to electron and holes, D is the diffusion coefficient, n is the carrier concentration, σ the electrical conductivity, and E the electric field. For $J = 0$, $E(=E_{junct})$ is thus a finite quantity owing to the spatial dependence of carrier concentrations.

$$K_{z0} = \frac{n^2}{1 - \sqrt{3}\delta} \approx n^2(1 + \sqrt{3}\delta) \tag{3.63c}$$

The quantities ν_m and μ_m ($m = x, y, z$) are determined by the following six conditions:

$$K_m(-\infty) = K_1 \equiv n^2(1 - \Delta_1) \tag{3.64a}$$

$$K_m(\infty) = K_2 \equiv n^2(1 - \Delta_2) \tag{3.64b}$$

where $m = x, y, z$. Equations (3.64) imply that the media at $x \to \pm\infty$ is the undisturbed (without junction effects) medium. The above conditions yield

$$\nu_m = K_{m0} - [(K_{m0} - K_1)(K_{m0} - K_2)]^{1/2} \tag{3.65a}$$

$$\mu_m = \tfrac{1}{4} \ln[(K_{m0} - K_2)/(K_{m0} - K_1)] \tag{3.65b}$$

These parameters are real if

$$K_{m0} > K_1 \qquad K_{m0} > K_2$$

If $\Delta_1 = \Delta_2$ or $K_1 = K_2 = K_0$, i.e., the dielectric constants are equal on the n and p sides of the junction, we will have

$$\nu_m = K_0 \qquad \mu_m = 0$$

and

$$K_m(x) = K_0 + \frac{K_{m0} - K_0}{\cosh^2(x/a)} \tag{3.66}$$

In Fig. 3.5 we have plotted a typical x dependence of two of the principal dielectric constants.

The modes can again be classified as either TE or TM. For TE modes, we may write $E_x = E_z = 0$ and with

$$\mathscr{E}_y = E_y(x) \exp[i(\omega t - \beta z)]$$

we obtain

$$\frac{d^2 E_y}{dx^2} + [k_0^2 K_y(x) - \beta^2] E_y(x) = 0 \tag{3.67}$$

where k_0 is the free space wave number. Substituting

$$\xi = \frac{x - \mu_y a}{a} \tag{3.68}$$

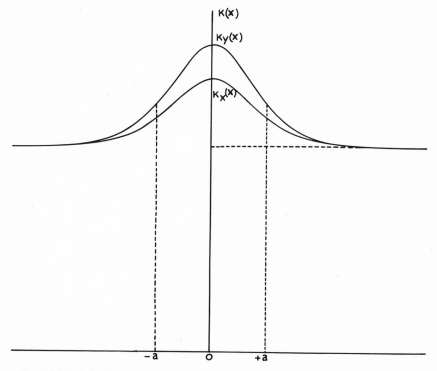

Fig. 3.5. Typical x dependence of two of the principal dielectric constants [as given by Eq. (3.66) for $m = x, y$] in the continuous dielectric constant model (after Nelson and McKenna, 1967; reprinted by permission).

in Eq. (3.67) we obtain

$$\frac{d^2 E_y}{d\xi^2} + [e_y - v_y \cosh 2\mu_y - v_y \sinh 2\mu_y \tanh \xi$$

$$+ v_y \cosh^2 \mu_y \operatorname{sech}^2 \xi] E_y(\xi) = 0 \qquad (3.69)$$

where

$$e_y = (k_0 a)^2 K_{y0} - (a\beta)^2 \qquad (3.70a)$$

and

$$v_y = (k_0 a)^2 (K_{y0} - \nu_y) \qquad (3.70b)$$

Similarly, for TM modes $H_x = H_z = 0$ and if we write

$$\mathcal{H}_y = H_y(x) \exp[i(\omega t - \beta z)]$$

then $H_y(x)$ would satisfy the following differential equation

$$\frac{d^2 H_y}{dx^2} - \frac{1}{K_z(x)} \frac{dK_z}{dx} \frac{dH_y}{dx} + \frac{K_z(x)}{K_x(x)} [k_0^2 K_x(x) - \beta^2] H_y(x) = 0 \qquad (3.71)$$

It can be shown that for realistic values of various parameters, only a small error will be involved if the first derivative term is neglected and if we set $K_z(x)/K_x(x) = 1$. This approximation will reduce Eq. (3.71) to the same form as Eq. (3.67). As for TE modes, if we make the substitution

$$\eta = \frac{x - \mu_x a}{a} \qquad (3.72)$$

we would obtain

$$\frac{d^2 H_y}{d\eta^2} + [e_x - v_x \cosh 2\mu_x - v_x \sinh 2\mu_x \tanh \eta$$

$$+ v_x \cosh^2 \mu_x \operatorname{sech}^2 \eta] H_y = 0 \qquad (3.73)$$

where

$$e_x = (k_0 a)^2 K_{x0} - (a\beta)^2 \qquad (3.74a)$$

and

$$v_x = (k_0 a)^2 (K_{x0} - v_x) \qquad (3.74b)$$

Equations (3.69) and (3.73) are of the same form as those obtained in the Eckart model in quantum mechanics and exact solutions can be obtained (see, for example, Morse and Feshbach, 1953, p. 1650). We will quote the relevant results here. First, a necessary condition for the existence of the discrete modes of Eq. (3.69) is given by

$$v_y > [\exp(2\mu_y)] \tanh \mu_y$$

or

$$(k_0 a)^2 (K_{y0} - K_1) > \frac{[(K_{y0} - K_2)^{1/2} - (K_{y0} - K_1)^{1/2}]}{[(K_{y0} - K_2)^{1/2} + (K_{y0} - K_1)^{1/2}]} \qquad (3.75)$$

Similarly for discrete modes of Eq. (3.73) to exist we must have

$$v_x > [\exp(2\mu_x)] \tanh \mu_x$$

or

$$(k_0 a)^2 (K_{x0} - K_1) > \frac{[(K_{x0} - K_2)^{1/2} - (K_{x0} - K_1)^{1/2}]}{[(K_{x0} - K_2)^{1/2} + (K_{x0} - K_1)^{1/2}]} \qquad (3.76)$$

The propagation constants, β_{N_m}, are determined from the relation

$$e_m[=(k_0a)^2K_{m0}-(a\beta)^2]$$
$$= v_m \cosh 2\mu_m - [(v_m \cosh^2 \mu_m + \tfrac{1}{4})^{1/2} - (N_m + \tfrac{1}{2})]^2$$
$$+ \frac{\tfrac{1}{4}(v_m \sinh 2\mu_m)^2}{[(v_m \cosh^2 \mu_m + \tfrac{1}{4})^{1/2} - (N_m + \tfrac{1}{2})]^2} \tag{3.77}$$

where $m = x, y$ and N_m takes on the finite sequence of integer values

$$N_m = 0, 1, 2, \ldots, < [(v_m \cosh^2 \mu_m + \tfrac{1}{4})^{1/2} - \tfrac{1}{2} - (\tfrac{1}{2}v_m \sinh 2\mu_m)^{1/2}] \tag{3.78}$$

If one assumes typical values for a GaP electrooptic modulator with $a = 0.3 \ \mu\text{m}$, $\lambda = 0.633 \ \mu\text{m}$, $n = 3.31$, and $\Delta \approx 10^{-3}$ then Nelson and Mc-Kenna (1967) have shown that only the zero order mode, $N_m = 0$ $(m = x, y)$, is possible.

3.4. Summary

In this chapter we have presented the modal analysis of guided propagation in inhomogeneous planar waveguides; the specific forms of spatial dependence of dielectric constants which have been analyzed are those which correspond to realistic situations and also lead to solutions in the closed form. However, there are many other physically significant forms of the spatial dependence of the dielectric constant for which solutions in a closed form employing modal analysis are not possible. Such cases have to be analyzed by approximate methods which have been developed in Chapter 6.

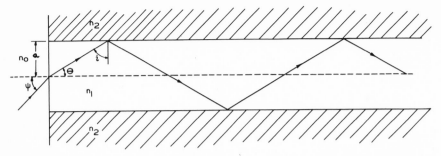

Fig. 4.1. The optical fiber; *i* denotes the angle of incidence of a ray incident on the fiber.

(ii) The number of guided modes is determined (as will be shown later) by n_1, n_2, and the core radius. Thus, for a given core radius and core material, the number of guided modes depends very much on the value of n_2. Often, it is desirable to have single-mode propagation; i.e., there exists only one mode which propagates. For such cases it is necessary to have suitable cladding material.

Fabrication of optical fibers has been extensively reviewed at many places and excellent books are now available which discuss the manufacture and mechanical properties very thoroughly (see, for example, Allan, 1973). As such we will not go into the fabricating techniques and would rather concentrate on the propagation characteristics of such a fiber.

We will first present propagation of rays through optical fibers, which will be followed by an approximate modal analysis based on the assumption that the modes are roughly transverse in character. Finally, a rigorous modal analysis will be presented.

4.1. The Ray Analysis

As mentioned earlier, for guided propagation the rays must undergo total internal reflection at the core–cladding interface.* Consider a ray which is incident on the entrance aperture of the guide making an angle *i* with the axis (see Fig. 4.1). Let the refracted ray make an angle θ with the normal to the core–cladding interface. Assuming the outside medium to have a refractive index n_0 (which for most practical cases is unity), we get

$$\frac{\sin i}{\sin[(\pi/2)-\theta]} = \frac{n_1}{n_0}$$

*We will consider propagation of meridional rays, i.e., rays whose paths are confined to a single plane.

or

$$\cos \theta = (n_0/n_1) \sin i \tag{4.2}$$

Further, for total internal reflection to occur at the core-cladding interface we must have

$$\sin \theta > n_2/n_1$$

or

$$\cos \theta < [1 - (n_2/n_1)^2]^{1/2} \tag{4.3}$$

Thus, the angle of incidence should satisfy the inequality

$$\sin i < \frac{n_1}{n_0}\left[1 - \left(\frac{n_2}{n_1}\right)^2\right]^{1/2} = \frac{(n_1^2 - n_2^2)^{1/2}}{n_0}$$

It is interesting to note that if $(n_1^2 - n_2^2)^{1/2} > n_0$ (or, $n_1^2 > n_2^2 + n_0^2$), then for all values of i, total internal reflection will occur. Assuming $n_0 = 1$, the maximum value of i, i_m, would be given by

$$\begin{aligned} \sin i_m &= (n_1^2 - n_2^2)^{1/2} &\quad \text{when } n_1^2 < n_2^2 + 1 \\ &= 1 &\quad \text{when } n_1^2 \geqslant n_2^2 + 1 \end{aligned} \tag{4.4}$$

Thus, if a cone of light is incident on one end of the fiber it will be guided through it, provided the semiangle of the cone is less than i_m. This angle is a measure of the light gathering power, and consequently one defines the numerical aperture (N.A.) of the fiber by the following equation

$$\text{N.A.} = (n_1^2 - n_2^2)^{1/2} \tag{4.5}$$

Although the above analysis is valid only for meridional rays, Eq. (4.4) is reasonably accurate even for skew rays. The effect of angled end faces has been discussed by Allan (1973).

4.2. Solutions of Bessel's Equation

In the following two sections we will solve Maxwell's equations to determine the modes of the optical fiber. However, before we carry out the mathematical analysis it is necessary to study the solutions of Bessel's equation

$$x^2 \frac{d^2 y}{dx^2} + x \frac{dy}{dx} + (x^2 - l^2) y = 0 \tag{4.6}$$

For nonintegral values of l the two linearly independent solutions are

$$J_l(x)$$

and

$$J_{-l}(x) \tag{4.7}$$

where

$$J_l(x) = \sum_{r=0}^{\infty} \frac{(-1)^r}{r!\,\Gamma(l+r+1)}\left(\frac{x}{2}\right)^{l+2r} \tag{4.8}$$

For $l = 0, 1, 2, \ldots$, $J_{-l}(x)$ is a multiple of $J_l(x)$:

$$J_{-l}(x) = (-1)^l J_l(x) \qquad l = 0, 1, 2, \ldots \tag{4.9}$$

For such a case

$$Y_l(x) = \lim_{\nu \to l} \frac{\cos \nu\pi[J_\nu(x) - J_{-\nu}(x)]}{\sin \nu\pi} \tag{4.10}$$

is the other independent solution. In fact $Y_l(x)$ (which is known as the Weber's function) is always a second independent solution. The asymptotic forms are given below:

(i) For $x \to 0$

$$J_l(x) \sim \frac{1}{\Gamma(l+1)}\left(\frac{x}{2}\right)^l \qquad \text{for all } l \tag{4.11}$$

$$Y_0(x) \sim \frac{2}{\pi} \ln x \tag{4.12}$$

$$Y_l(x) \sim -\frac{\Gamma(l)}{\pi}\left(\frac{2}{x}\right)^l \qquad \text{for } l > 0 \tag{4.13}$$

(ii) For $x \to \infty$

$$J_l(x) \sim \left(\frac{2}{\pi x}\right)^{1/2} \cos\left(x - \frac{l\pi}{2} - \frac{\pi}{4}\right) \tag{4.14}$$

$$Y_l(x) \sim \left(\frac{2}{\pi x}\right)^{1/2} \sin\left(x - \frac{l\pi}{2} - \frac{\pi}{4}\right) \tag{4.15}$$

It should be seen that for small values of x, $Y_l(x)$ tends to $-\infty$.
We next consider the equation

$$x^2\frac{d^2y}{dx^2} + x\frac{dy}{dx} - (x^2 + l^2)y = 0 \tag{4.16}$$

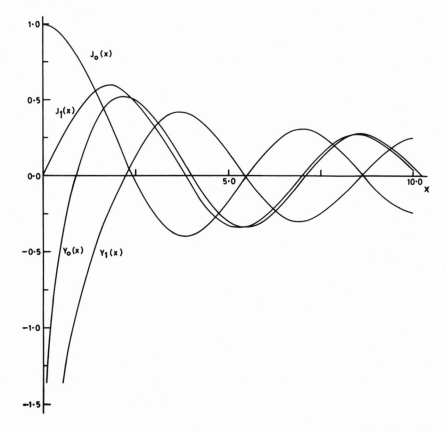

Fig. 4.2. Variation of $J_0(x)$, $J_1(x)$, $Y_0(x)$, and $Y_1(x)$ with x. Notice that both Y_0 and Y_1 tend to $-\infty$ as $x \to 0$.

the solutions of which can easily be seen to be $J_l(ix)$ and $Y_l(ix)$, where $i = \sqrt{-1}$. However, it is convenient to define real solutions in the following way:

$$I_l(x) = i^{-l} J_l(ix) = \sum_{r=0}^{\infty} \frac{1}{r!\Gamma(l+r+1)} \left(\frac{x}{2}\right)^{l+2r} \qquad (4.17)$$

whose behavior is not oscillatory and tends to ∞ as $x \to \infty$. The other independent solution is defined by

$$K_l(x) = \lim_{\nu \to l} \frac{\pi}{2} \frac{I_{-\nu}(x) - I_{\nu}(x)}{\sin \nu\pi} \qquad (4.18)$$

which tend to ∞ as $x \to 0$. The asymptotic forms are given below:

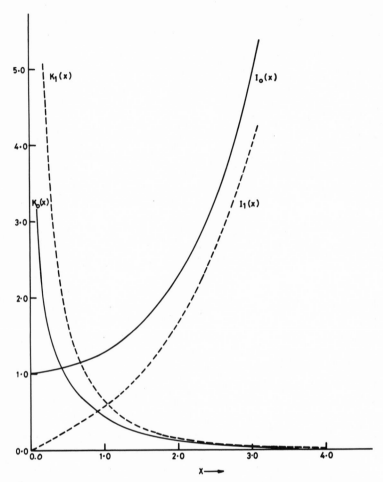

Fig. 4.3. Variation of $K_0(x)$, $K_1(x)$, $I_0(x)$, and $I_1(x)$. Notice that the K functions are monotonically decreasing functions of x and tend to ∞ as $x \to 0$; whereas the I functions are monotonically increasing functions of x and tend to ∞ as $x \to \infty$.

(i) For $x \to 0$

$$I_l(x) \sim \frac{1}{\Gamma(l+1)}\left(\frac{x}{2}\right)^l \qquad \text{for all } l \qquad (4.19)$$

$$K_0(x) \sim -I_0(x) \ln (x/2) \qquad (4.20)$$

$$K_l(x) \sim \tfrac{1}{2}\Gamma(l)(2/x)^l \qquad \text{for } l > 0 \qquad (4.21)$$

(ii) For $x \to \infty$

$$I_l(x) \sim \frac{1}{(2\pi x)^{1/2}} e^x \tag{4.22}$$

$$K_l(x) \sim \left(\frac{\pi}{2x}\right)^{1/2} e^{-x} \tag{4.23}$$

Sketches of $J_l(x)$ and $Y_l(x)$ are given in Fig. 4.2 and of $I_l(x)$ and $K_l(x)$ are given in Fig. 4.3.

4.3. Modal Analysis (Approximate Theory)

In this section we will obtain solutions of Maxwell's equation which correspond to guided propagation in the optical fiber. We will follow the treatment of Gloge (1971a); this is based on the assumption

$$(n_1 - n_2)/n_1 \ll 1 \tag{4.24}$$

which is valid for most practical fibers. In the above equation n_1 and n_2 represent the core and cladding refractive indices. When Eq. (4.24) is valid we can construct modes whose transverse field is essentially polarized in one direction (we will show later that this is indeed a valid approximation). Since the refractive index is constant in each region, any cartesian component of **E** or **H** will satisfy the scalar wave equation. Further, since the system has cylindrical symmetry we may assume, without any loss of generality, that the transverse component of the electric field is along the y axis. This component, denoted by $\mathcal{E}_y$, will satisfy the following scalar wave equation

$$\nabla^2 \mathcal{E}_y = \varepsilon \mu_0 \frac{\partial^2 \mathcal{E}_y}{\partial t^2} \tag{4.25}$$

where

$$\begin{aligned}
\varepsilon = \varepsilon_1 &= \varepsilon_0 n_1^2 && \text{for } 0 < r < a \\
&= \varepsilon_2 = \varepsilon_0 n_2^2 && \text{for } r > a
\end{aligned} \tag{4.26}$$

ε_0 being the free space permittivity. For waves propagating in the z direction, the z and t dependence of the field would be of the form

$$\mathcal{E}_y = E_y \exp[i(\omega t - \beta z)] \tag{4.27}$$

Thus

$$\nabla_\perp^2 E_y + (n^2 k_0^2 - \beta^2) E_y = 0 \tag{4.28}$$

where

$$\nabla_{\perp}^2 \equiv \nabla^2 - \frac{\partial^2}{\partial z^2} \tag{4.29}$$

and

$$k_0 = \omega(\varepsilon_0\mu_0)^{1/2} = \omega/c$$

is the free space wave number. In the cylindrical system of coordinates we obtain

$$\frac{\partial^2 E_y}{\partial r^2} + \frac{1}{r} \frac{\partial E_y}{\partial r} + \frac{1}{r^2} \frac{\partial^2 E_y}{\partial \varphi^2} + (n^2 k_0^2 - \beta^2)E_y = 0 \tag{4.30}$$

Writing

$$E_y = R(r)\Phi(\varphi)$$

we obtain

$$\frac{1}{R(r)}\left[r^2\frac{d^2R}{dr^2} + r\frac{dR}{dr} + \{n^2(r)k_0^2 - \beta^2\}r^2R(r)\right] = -\frac{1}{\Phi}\frac{d^2\Phi}{d\varphi^2}$$
$$= l^2 \tag{4.31}$$

Since the LHS is a function of r alone and the RHS is a function of φ alone, each side has been set equal to a constant $(=l^2)$. Thus $\Phi(\varphi)$ must be a linear combination of $\cos l\varphi$ and $\sin l\varphi$. Further, for E_y to be single valued $[E_y(\varphi + 2\pi) = E_y(\varphi)]$, l must be 0, 1, 2, ... Further

$$r^2\frac{d^2R}{dr^2} + r\frac{dR}{dr} + [(n^2k_0^2 - \beta^2)r^2 - l^2]R(r) = 0 \tag{4.32}$$

First consider the region $r > a$, where $n = n_2$. Clearly β^2 should be greater than $n_2^2k_0^2$ because only in this case will we have an equation of the form (4.16) and the fields will exponentially decay in the cladding [in fact we must choose the K_l functions as our solution; see Eqs. (4.22) and (4.23)]. For $\beta^2 < n_1^2k_0^2$, the solution will be in terms of Bessel and Weber functions which do not decay exponentially for large values of r [see Eqs. (4.14) and (4.15)].*

*It should be pointed out that there are two kinds of modes (i) guided modes and (ii) radiation modes. The guided modes are characterized by the property that they tend to zero exponentially in the limit of $r \to \infty$. On the other hand the radiation modes do not decay exponentially as $r \to \infty$, i.e., the fiber acts more like an antenna than a waveguide. For the waveguiding properties we would be interested only in the guided modes, although when the fiber geometry is not perfect the mode conversion that results not only mixes the guided modes but also the radiation modes (see Chapter 8). Another difference between these two is the fact that the guided modes are discrete while the radiation modes form a continuum. (The metallic waveguides differ from the dielectric waveguides in the absence of these radiation modes, since the fields have to satisfy the boundary condition of going to zero at the walls of the waveguide.) In this chapter we will be considering only the guided modes which will be characterized by an exponential decay in the cladding.

Further, the quantity $(n^2k_0^2-\beta^2)$ should not be negative everywhere, because then in the region $0<r<a$ we must choose I_l as our solution (otherwise the field will be infinite at the origin; see Fig. 4.3) and in the region $r>a$ we must choose K_l as our solution (otherwise the field will not decay in the cladding); the I_l functions are monotonically increasing and the K_l functions are monotonically decreasing and one can never match the derivative at $r=a$. Thus $(n^2k_0^2-\beta^2)$ must be positive in the region $0<r<a$ and therefore

$$\beta^2<n_1^2k_0^2 \tag{4.33}$$

Thus, we must have the inequality

$$n_2^2k_0^2<\beta^2<n_1^2k_0^2 \tag{4.34}$$

The above inequality should also have been qualitatively obvious from the fact that when a (the radius of the core) tends to ∞ we essentially have an infinitely extended homogeneous medium with refractive index n_1 and therefore β should be equal to n_1k_0. On the other hand, when $a\rightarrow 0$ the medium corresponds to refractive index n_2 and $\beta=n_2k_0$. For a finite value of a we should expect β to lie between n_2k_0 and n_1k_0.

We define three real dimensionless parameters

$$u=a(n_1^2k_0^2-\beta^2)^{1/2} \tag{4.35}$$

$$w=a(\beta^2-n_2^2k_0^2)^{1/2} \tag{4.36}$$

$$v=(u^2+w^2)^{1/2}$$
$$=ak_0(n_1^2-n_2^2)^{1/2} \tag{4.37}$$

where a represents the radius of the core. The parameter v is often referred to as the normalized frequency. In terms of these parameters Eq. (4.32) can be written in the form

$$r^2\frac{d^2R}{dr^2}+r\frac{dR}{dr}+\left(\frac{u^2}{a^2}r^2-l^2\right)R(r)=0 \qquad \text{for } 0<r<a \tag{4.38}$$

and

$$r^2\frac{d^2R}{dr^2}+r\frac{dR}{dr}-\left(\frac{w^2}{a^2}r^2+l^2\right)R(r)=0 \qquad \text{for } r>a \tag{4.39}$$

Clearly, for the solutions to be bounded we must choose $J_l(ur/a)$ as the solution for $0<r<a$ and $K_l(wr/a)$ as the solution for $r>a$. Thus we obtain

$$E_y=E_l\frac{1}{J_l(u)}J_l\left(\frac{ur}{a}\right)\cos l\varphi \qquad \text{for } 0<r<a$$

$$=E_l\frac{1}{K_l(w)}K_l\left(\frac{wr}{a}\right)\cos l\varphi \qquad \text{for } r>a \tag{4.40}$$

where we have assumed the continuity of E_y at the interface*; E_l represents the electric field at the interface $(r = a)$. Since we have the freedom of choosing $\sin l\varphi$ or $\cos l\varphi$ in Eq. (4.40) and two orthogonal states of polarization, we can construct a set of four modes for all $l > 0$. For $l = 0$ we have only two modes corresponding to the two states of polarization. Since

$$\nabla \times \mathscr{E} = -\frac{\partial \mathscr{B}}{\partial t} = -\mu_0 \frac{\partial \mathscr{H}}{\partial t} \tag{4.41}$$

we have, for approximate plane waves

$$-i\omega\mu_0 H_x \approx +i\beta E_y \approx +ink_0 E_y \tag{4.42}$$

or

$$E_y = -\frac{\omega\mu_0}{k_0 n} H_x = -\frac{Z_0}{n} H_x$$

where n could be either n_1 or n_2, and $Z_0[=(\omega\mu_0/k_0) = c\mu_0 = (1/c\varepsilon_0)]$ is known as the plane wave impedance in vacuum. Similarly

$$-i\omega\mu_0 H_z = \frac{\partial E_y}{\partial x}$$

or,

$$H_z = \frac{i}{\omega\mu_0} \frac{\partial E_y}{\partial x} = \frac{i}{k_0 Z_0} \frac{\partial E_y}{\partial x} \tag{4.43}$$

Further, the z component of the equation

$$\nabla \times \mathscr{H} = \frac{\partial \mathscr{D}}{\partial t} \tag{4.44}$$

gives

$$E_z = \frac{i}{\omega\varepsilon} \frac{\partial H_x}{\partial y}$$

or

$$E_z = \frac{iZ_0}{k_0 n^2} \frac{\partial H_x}{\partial y} = -\frac{i}{nk_0} \frac{\partial E_y}{\partial y} \tag{4.45}$$

*This is not rigorously correct because one must make E_φ, E_z, and εE_r to be continuous at the interface. The error involved is negligible as long as the inequality expressed by Eq. (4.24) is satisfied. We may easily see this by noting that at $\varphi = 0$, E_y (which is a tangential component at $\varphi = 0$) should be continuous. On the other hand, at $\varphi = \pi/2$, εE_y should be continuous. Continuity of E_y at all values of φ is consistent with our approximation that $n_1 \approx n_2$.

Equations (4.43) and (4.45) give the longitudinal components. In order to calculate E_z and H_z we note that

$$\frac{\partial E_y}{\partial y} = \frac{\partial E_y}{\partial r}\frac{\partial r}{\partial y} + \frac{\partial E_y}{\partial \varphi}\frac{\partial \varphi}{\partial y}$$

$$= \frac{\partial E_y}{\partial r}\sin \varphi + \frac{1}{r}\frac{\partial E_y}{\partial \varphi}\cos \varphi$$

Thus

$$\frac{\partial E_y}{\partial y} = \frac{E_l}{J_l(u)}\left[\frac{u}{a}J_l'(\xi)\sin \varphi \cos l\varphi - \frac{l}{r}J_l(\xi)\sin l\varphi \cos \varphi\right] \tag{4.46}$$

where $\xi = ur/a$ and primes denote differentiation with respect to the argument. Now

$$J_l'(\xi) = \tfrac{1}{2}[J_{l-1}(\xi) - J_{l+1}(\xi)] \tag{4.47}$$

and

$$J_l(\xi) = (\xi/2l)[J_{l-1}(\xi) + J_{l+1}(\xi)] \tag{4.48}$$

On substituting the above expressions for $J_l'(\xi)$ and $J_l(\xi)$ in Eq. (4.46) and carrying out simple manipulations, we obtain

$$E_z = \frac{iE_l}{2k_0 a}\left[\frac{u}{n_1}\frac{J_{l+1}(ur/a)}{J_l(u)}\sin(l+1)\varphi\right.$$

$$\left. + \frac{u}{n_1}\frac{J_{l-1}(ur/a)}{J_l(u)}\sin(l-1)\varphi\right] \qquad \text{for } 0 < r < a \quad (4.49a)$$

$$= \frac{iE_l}{2k_0 a}\left[\frac{w}{n_1}\frac{K_{l+1}(wr/a)}{K_l(w)}\sin(l+1)\varphi\right.$$

$$\left. - \frac{w}{n_1}\frac{K_{l-1}(wr/a)}{K_l(w)}\sin(l-1)\varphi\right] \qquad \text{for } r > a \quad (4.49b)$$

Similarly

$$H_z = -\frac{iE_l}{2k_0 Z_0 a}\left[u\frac{J_{l+1}(ur/a)}{J_l(u)}\cos(l+1)\varphi\right.$$

$$\left. - u\frac{J_{l-1}(ur/a)}{J_l(u)}\cos(l-1)\varphi\right] \qquad \text{for } 0 < r < a \quad (4.50a)$$

$$= -\frac{iE_l}{2k_0 Z_0 a}\left[w\frac{K_{l+1}(wr/a)}{K_l(w)}\cos(l+1)\varphi\right.$$

$$\left. + w\frac{K_{l-1}(wr/a)}{K_l(w)}\cos(l-1)\varphi\right] \qquad \text{for } r > a \quad (4.50b)$$

The strengths of the longitudinal components are small compared to the strengths of the transverse components. This can easily be seen in the following way:

$$\left|\frac{E_z}{E_y}\right| \sim \frac{u}{ak_0} = \frac{(k_0^2 n_1^2 - \beta^2)^{1/2}}{k_0} < (n_1^2 - n_2^2)^{1/2} \sim \sqrt{\Delta} \qquad (4.51)$$

and a similar equation for the magnetic field components. Thus, as long as $\Delta \ll 1$, the present theory is expected to give fairly accurate results.

Continuity of E_z (which is a tangential component) at $r = a$ gives

$$u\frac{J_{l+1}(u)}{J_l(u)}\sin(l+1)\varphi + u\frac{J_{l-1}(u)}{J_l(u)}\sin(l-1)\varphi$$

$$= w\frac{K_{l+1}(w)}{K_l(w)}\sin(l+1)\varphi - w\frac{K_{l-1}(w)}{K_l(w)}\sin(l-1)\varphi \qquad (4.52)$$

where we have set $n_1 \approx n_2$ which is consistent with our approximation scheme. Using the relations

$$J_{l+1}(u) = (2l/u)J_l(u) - J_{l-1}(u) \qquad (4.53)$$

and

$$K_{l+1}(w) = (2l/w)K_l(w) + K_{l-1}(w) \qquad (4.54)$$

we get

$$2l\sin(l+1)\varphi - u\frac{J_{l-1}(u)}{J_l(u)}[\sin(l+1)\varphi - \sin(l-1)\varphi]$$

$$= 2l\sin(l+1)\varphi + w\frac{K_{l-1}(w)}{K_l(w)}[\sin(l+1)\varphi - \sin(l-1)\varphi]$$

which gives

$$u\frac{J_{l-1}(u)}{J_l(u)} = -w\frac{K_{l-1}(w)}{K_l(w)} \qquad (4.55)$$

Equation (4.55) is known as the characteristic equation for the linearly polarized (LP) modes. Continuity of H_z would have also led to the same characteristic equation. For $l = 0$ we get

$$u\frac{J_1(u)}{J_0(u)} = w\frac{K_1(w)}{K_0(w)}$$

where we have used the relation $K_{-n}(x) = K_n(x)$.

The φ components of the field, E_φ and H_φ, can also be calculated from Eqs. (4.40) and (4.42):

$$E_\varphi = E_y \cos\varphi = \tfrac{1}{2}E_l \frac{J_l(ur/a)}{J_l(u)}$$

$$\times [\cos(l+1)\varphi + \cos(l-1)\varphi] \qquad \text{for } 0<r<a$$

$$= \tfrac{1}{2}E_l \frac{K_l(wr/a)}{K_l(w)}$$

$$\times [\cos(l+1)\varphi + \cos(l-1)\varphi] \qquad \text{for } r>a$$

(4.56)

and

$$H_\varphi = H_x \sin\varphi = -\tfrac{1}{2}\frac{E_l}{Z_0}n_1\frac{J_l(ur/a)}{J_l(u)}$$

$$\times [\sin(l+1)\varphi - \sin(l-1)\varphi] \qquad \text{for } 0<r<a$$

$$= -\tfrac{1}{2}\frac{E_l}{Z_0}n_2\frac{K_l(wr/a)}{K_l(w)}$$

$$\times [\sin(l+1)\varphi - \sin(l-1)\varphi] \qquad \text{for } r>a$$

(4.57)

If we set $n_1 = n_2$, both E_φ and H_φ are also continuous at $r = a$.

For given values of a, n_1, and n_2, Eq. (4.55) will determine the allowed values of β. The cutoff values correspond to $\beta = n_2 k_0$ and therefore $w = 0$. Thus we must have

$$J_{l-1}(u) = 0 \qquad\qquad (4.58)$$

For $l = 0$ Eq. (4.58) includes the roots of $J_{-1}(u)$ $[= -J_1(u)]$ which (following Gloge, 1971a) we shall count so as to include $J_1(0) = 0$ as the first root. Since the zeros of $J_1(x)$ occur at $x = 0$, 3.8317, 7.0156, and so on, and of $J_0(x)$ occur at $x = 2.4048$, 5.5201, 8.6537, and so on, the cutoff values for the LP_{01}, LP_{02}, and LP_{03} modes* occur at $u = 0$, 3.8317, and 7.0156, respectively, and the cutoff values for the LP_{11}, LP_{12}, and LP_{13} occur at $u = 2.4048$, 5.5201, and 8.6537, respectively (see Fig. 4.4 where the curves intersect the horizontal axis at these points).

In the limit of large w, Eq. (4.55) gives

$$J_l(u) = 0$$

Thus, the solutions for u are between the zeros of $J_{l-1}(u)$ and $J_l(u)$. For example, the u value corresponding to the LP_{01} mode has to lie between 0

---

*The first subscript on LP refers to the l value and the second subscript corresponds to the various roots for that particular l value.

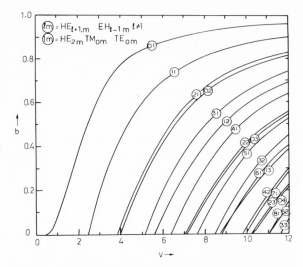

Fig. 4.4. Normalized propagation parameter $b = [(\beta/k) - n_2]/(n_1 - n_2)$ as a function of the normalized frequency v (after Gloge, 1971a; reprinted by permission).

and 2.4048; similarly for the LP_{12} mode, u must lie between 5.5201 and 7.0156 (see Figs. 4.4 and 4.5).

As mentioned before, for $l \geqslant 1$, each set (designated as LP_{lm}) comprises four modes which are degenerate. If we carry out an accurate analysis, then each mode LP_{lm} can be shown to break up into modes which can be identified as $HE_{l+1,m}$ and $EH_{l-1,m}$, or TE_{0m} and TM_{0m} modes. A rigorous analysis is given in the next section; however, the present analysis has been found to be reasonably accurate for most practical systems.

Looking back at Fig. 4.4 we note that the LP_{01} mode (which is known as the HE_{11} mode) has zero cutoff frequency and no matter what the values of a, n_1, and n_2, this mode will always propagate. Further, if the values of a, n_1, and n_2 are such that

$$0 < v < 2.4048 \tag{4.59}$$

only the HE_{11} mode will propagate. Such single mode operation is of great interest, since under multimode operation different modes propagate with different velocities, which causes considerable signal distortion over long distances. This is discussed in Chapter 7 where we have also calculated the group velocity of various modes.

Gloge (1971a) has also obtained an approximate analytic solution of Eq. (4.55) when v is large (see also Snyder, 1969). We merely give the results. For all modes excluding the HE_{11} mode

$$u(v) \approx \frac{u_c \exp[\sin^{-1}(s/u_c) - \sin^{-1}(s/v)]}{s} \qquad (4.60)$$

where

$$s = (u_c^2 - l^2 - 1)^{1/2}$$

and u_c is the mth root of $J_{l-1}(u)$ (we are considering the LP_{lm} mode). For the HE_{11} mode

$$u(v) \approx \frac{(1+\sqrt{2})v}{1 + (4 + v^2)^{1/4}} \qquad (4.61)$$

When $v \gg s$ (i.e., far from cutoff), Eqs. (4.60) and (4.61) reduce to

$$u(v) \approx u_\infty[1 - (1/v)] \qquad (4.62)$$

where u_∞ is the mth root of $J_l(u)$.

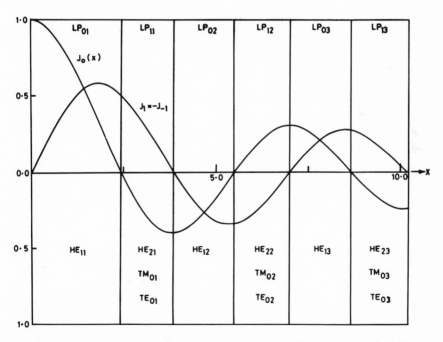

Fig. 4.5. Regions of the parameter u for modes corresponding to $l = 0$ and $l = 1$ (after Gloge, 1971a; reprinted by permission).

4.3.1.   The Poynting Vector

The Poynting vector has been defined in Sec. 2.3. For the present case it would be given by

$$\bar{S}_z \approx \frac{nk_0}{2\omega\mu_0}E_y^2 \approx \frac{n}{2Z_0}E_y^2$$

Thus

$$\bar{S}_z \approx \frac{1}{2}\frac{n_1}{Z_0}E_l^2\left[\frac{J_l(ur/a)}{J_l(u)}\right]^2 \cos^2 l\varphi \qquad \text{for } 0 < r < a$$

$$\approx \frac{1}{2}\frac{n_2}{Z_0}E_l^2\left[\frac{K_l(wr/a)}{K_l(w)}\right]^2 \cos^2 l\varphi \qquad \text{for } r > a \tag{4.63}$$

Thus the power contained in the core is given by

$$P_{core} = \int_0^a \int_0^{2\pi} \bar{S}_z r\, dr\, d\varphi \tag{4.64}$$

$$= A_l \frac{\pi a^2}{2}\frac{n_1}{Z_0}E_l^2 \frac{1}{[uJ_l(u)]^2}\int_0^u J_l^2(\xi)\xi\, d\xi$$

$$= A_l \frac{\pi a^2}{4}\frac{n_1}{Z_0}E_l^2\left\{1 - \frac{J_{l-1}(u)J_{l+1}(u)}{[J_l(u)]^2}\right\} \tag{4.65}$$

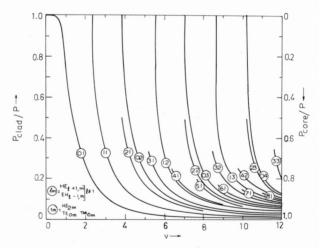

Fig. 4.6. Variation of the fractional power contained in the cladding P_{clad}/P ($= 1 - P_{core}/P$) with v (after Gloge, 1971a; reprinted by permission).

where

$$A_l = 2 \qquad \text{for } l = 0$$
$$= 1 \qquad \text{for } l = 1, 2, \ldots$$

In a similar manner one can calculate the power contained in the cladding, given by

$$P_{\text{clad}} = \int_a^\infty \int_0^{2\pi} \bar{S}_z r \, dr \, d\varphi \qquad (4.66)$$

where the functions are once again integrable. The fractional powers contained in the core and in the cladding P_{core}/P and P_{clad}/P ($P = P_{\text{core}} + P_{\text{clad}}$) are plotted in Fig. 4.6. We notice that the power associated with a particular mode is concentrated in the core for large values of v (i.e., far away from the cutoff).

4.4. Rigorous Modal Analysis

In the previous section we have given an approximate modal analysis which is valid when $\Delta \ll 1$, i.e., when the refractive index of the cladding is very close to that of the core. We will carry out a rigorous analysis here and will compare the results with the ones obtained by using the approximate approach.

We will assume the field components to be of the form

$$\mathscr{E}_i(r, \varphi, z; t) = E_i(r) \exp[i(\omega t - \beta z + l\varphi)] \qquad (4.67)$$

$$\mathscr{H}_i(r, \varphi, z; t) = H_i(r) \exp[i(\omega t - \beta z + l\varphi)] \qquad (4.68)$$

where the subscript i stands for any component (i.e., r, φ or z) of the fields and for consistency with the existing literature, we have assumed the φ dependence to be of the form $\exp(il\varphi)$. Once again for the fields to be single valued l must be $0, \pm 1, \pm 2, \ldots$.

We will first express the transverse field components (E_r, E_φ, H_r, and H_φ) in terms of the longitudinal components (E_z and H_z); next we will solve the wave equations for E_z and H_z; then we will apply the continuity conditions (at $r = a$) for the tangential components of the fields which will give us the characteristic equation.

We start with the curl equations, which in the cylindrical system of coordinates are given by

$$\frac{1}{r}\frac{\partial \mathscr{E}_z}{\partial \varphi} - \frac{\partial \mathscr{E}_\varphi}{\partial z} = -\mu_0 \frac{\partial \mathscr{H}_r}{\partial t} \tag{4.69}$$

$$\frac{\partial \mathscr{E}_r}{\partial z} - \frac{\partial \mathscr{E}_z}{\partial r} = -\mu_0 \frac{\partial \mathscr{H}_\varphi}{\partial t} \tag{4.70}$$

$$\frac{1}{r}\left[\frac{\partial}{\partial r}(r\mathscr{E}_\varphi) - \frac{\partial \mathscr{E}_r}{\partial \varphi}\right] = -\mu_0 \frac{\partial \mathscr{H}_z}{\partial t} \tag{4.71}$$

$$\frac{1}{r}\frac{\partial \mathscr{H}_z}{\partial \varphi} - \frac{\partial \mathscr{H}_\varphi}{\partial z} = \varepsilon \frac{\partial \mathscr{E}_r}{\partial t} \tag{4.72}$$

$$\frac{\partial \mathscr{H}_r}{\partial z} - \frac{\partial \mathscr{H}_z}{\partial r} = \varepsilon \frac{\partial \mathscr{E}_\varphi}{\partial t} \tag{4.73}$$

$$\frac{1}{r}\left[\frac{\partial}{\partial r}(r\mathscr{H}_\varphi) - \frac{\partial \mathscr{H}_r}{\partial \varphi}\right] = \varepsilon \frac{\partial \mathscr{E}_z}{\partial t} \tag{4.74}$$

Using Eqs. (4.67) and (4.68), we get

$$\frac{il}{r}E_z + i\beta E_\varphi = -i\omega\mu_0 H_r \tag{4.75}$$

$$-i\beta E_r - \frac{d}{dr}E_z = -i\omega\mu_0 H_\varphi \tag{4.76}$$

$$\frac{1}{r}\left[\frac{d}{dr}(rE_\varphi) - ilE_r\right] = -i\omega\mu_0 H_z \tag{4.77}$$

$$\frac{il}{r}H_z + i\beta H_\varphi = i\omega\varepsilon E_r \tag{4.78}$$

$$-i\beta H_r - \frac{dH_z}{dr} = i\omega\varepsilon E_\varphi \tag{4.79}$$

$$\frac{1}{r}\left[\frac{d}{dr}(rH_\varphi) - ilH_r\right] = i\omega\varepsilon E_z \tag{4.80}$$

Equation (4.75) gives

$$H_r = \frac{i}{\omega\mu_0}\left(\frac{il}{r}E_z + i\beta E_\varphi\right)$$

On substitution in Eq. (4.79), we obtain

$$\frac{\beta}{\omega\mu_0}\left(\frac{il}{r}E_z + i\beta E_\varphi\right) - \frac{dH_z}{dr} = i\omega\varepsilon E_\varphi$$

Thus,

$$E_\varphi\left(i\omega\varepsilon - \frac{i\beta^2}{\omega\mu_0}\right) = \frac{il\beta}{\omega\mu_0}\frac{1}{r}E_z - \frac{dH_z}{dr}$$

or

$$E_\varphi = -\frac{i}{\kappa^2}\left[\frac{il\beta}{r}E_z - \omega\mu_0\frac{dH_z}{dr}\right] \tag{4.81}$$

where

$$\begin{aligned}\kappa^2 &= n^2\omega^2\varepsilon_0\mu_0 - \beta^2\\ &= n^2k_0^2 - \beta^2\end{aligned} \tag{4.82}$$

Similarly

$$E_r = -\frac{i}{\kappa^2}\left(\beta\frac{dE_z}{dr} + \omega\mu_0\frac{il}{r}H_z\right) \tag{4.83}$$

$$H_\varphi = -\frac{i}{\kappa^2}\left(\frac{il\beta}{r}H_z + \omega\varepsilon\frac{dE_z}{dr}\right) \tag{4.84}$$

and

$$H_r = -\frac{i}{\kappa^2}\left(\beta\frac{dH_z}{dr} - \omega\varepsilon\frac{il}{r}E_z\right) \tag{4.85}$$

Thus, if the longitudinal components of the electric and magnetic fields are known, the corresponding transverse components can be determined by using Eqs. (4.81)–(4.85). Since in each region (i.e., $0 < r < a$ and $r > a$) ε is a constant, the longitudinal components of the electric and magnetic fields would also satisfy the scalar wave equation [Eq. (4.30)]. Once again the propagation constants will satisfy the inequality (4.34) and E_z and H_z will be given by

$$\begin{aligned}E_z &= A\frac{1}{J_l(u)}J_l\left(\frac{ur}{a}\right) && \text{for } 0 < r < a\\ &= A\frac{1}{K_l(w)}K_l\left(\frac{wr}{a}\right) && \text{for } r > a\end{aligned} \tag{4.86}$$

and

$$\begin{aligned}H_z &= B\frac{1}{J_l(u)}J_l\left(\frac{ur}{a}\right) && \text{for } 0 < r < a\\ &= B\frac{1}{K_l(w)}K_l\left(\frac{wr}{a}\right) && \text{for } r > a\end{aligned} \tag{4.87}$$

where u and w are defined by Eqs. (4.35) and (4.36), and in writing Eqs. (4.86) and (4.87) we have incorporated the boundary condition that E_z and H_z are continuous at $r = a$. The other tangential components (E_φ and H_φ) should also be continuous and therefore we calculate E_φ and H_φ:

$$E_\varphi = -i\left(\frac{a}{u}\right)^2\left[A\frac{il\beta}{r}\frac{1}{J_l(u)}J_l\left(\frac{ur}{a}\right) - B\omega\mu_0\frac{1}{J_l(u)}J_l'\left(\frac{ur}{a}\right)\frac{u}{a}\right] \qquad \text{for } 0 < r < a$$

$$= i\left(\frac{a}{w}\right)^2\left[A\frac{il\beta}{r}\frac{1}{K_l(w)}K_l\left(\frac{wr}{a}\right) - B\omega\mu_0\frac{1}{K_l(w)}K_l'\left(\frac{wr}{a}\right)\frac{w}{a}\right] \qquad \text{for } r > a$$

$$(4.88)$$

and

$$H_\varphi = -i\left(\frac{a}{u}\right)^2\left[B\frac{il\beta}{r}\frac{1}{J_l(u)}J_l\left(\frac{ur}{a}\right) + A\omega\varepsilon_0 n_1^2\frac{1}{J_l(u)}J_l'\left(\frac{ur}{a}\right)\frac{u}{a}\right] \qquad \text{for } 0 < r < a$$

$$= i\left(\frac{a}{w}\right)^2\left[B\frac{il\beta}{r}\frac{1}{K_l(w)}K_l\left(\frac{wr}{a}\right) + A\omega\varepsilon_0 n_2^2\frac{1}{K_l(w)}K_l'\left(\frac{wr}{a}\right)\frac{w}{a}\right] \qquad \text{for } r > a$$

$$(4.89)$$

where primes denote differentiation with respect to the argument. Continuity of E_φ and H_φ at $r = a$ gives

$$A\frac{il\beta}{a}\left(\frac{1}{u^2} + \frac{1}{w^2}\right) - B\frac{\omega\mu_0}{a}\left[\frac{1}{u}\frac{J_l'(u)}{J_l(u)} + \frac{1}{w}\frac{K_l'(w)}{K_l(w)}\right] = 0 \qquad (4.90)$$

and

$$A\frac{\omega\varepsilon_0}{a}\left[\frac{n_1^2}{u}\frac{1}{J_l(u)}J_l'(u) + \frac{n_2^2}{w}\frac{1}{K_l(w)}K_l'(w)\right] + B\frac{il\beta}{a}\left(\frac{1}{u^2} + \frac{1}{w^2}\right) = 0 \qquad (4.91)$$

For nontrivial solutions the determinant must vanish and one obtains, after some simple manipulations,

$$\left[\frac{n_1^2}{n_2^2}\frac{w^2}{u}\frac{J_l'(u)}{J_l(u)} + w\frac{K_l'(w)}{K_l(w)}\right]\left[\frac{w^2}{u}\frac{J_l'(u)}{J_l(u)} + w\frac{K_l'(w)}{K_l(w)}\right] = \left[l\left(\frac{n_1^2}{n_2^2} - 1\right)\beta n_2 k_0\left(\frac{a}{u}\right)^2\right]^2 \qquad (4.92)$$

For given values of n_1, n_2, a, and ω the above equation can be solved for β corresponding to various values of l. We note that when n_1 is very close to n_2, the RHS is extremely small and one obtains approximately:

$$\frac{1}{u}\frac{J_l'(u)}{J_l(u)} \approx -\frac{1}{w}\frac{K_l'(w)}{K_l(w)} \qquad (4.93)$$

If one uses relations like

$$J_l'(u) = J_{l-1}(u) - \frac{l}{u} J_l(u) \tag{4.94}$$

one would obtain Eq. (4.55).

Next, let us consider the case $l = 0$. For this case it is worthwhile to go back to Eqs. (4.90) and (4.91). One obtains *either*

$$A = 0 \quad [\text{i.e., } E_z = 0]$$

and

$$\frac{1}{u} \frac{J_0'(u)}{J_0(u)} + \frac{1}{w} \frac{K_0'(w)}{K_0(w)} = 0 \tag{4.95}$$

or

$$B = 0 \quad [\text{i.e., } H_z = 0]$$

and

$$\frac{n_1^2}{n_2^2} \frac{1}{u} \frac{J_0'(u)}{J_0(u)} + \frac{1}{w} \frac{K_0'(w)}{K_0(w)} = 0 \tag{4.96}$$

Equations (4.95) and (4.96) correspond to TE and TM modes, respectively. For both cases the cutoff condition (i.e., $w \to 0$) corresponds to*

$$J_0(u) = 0 \tag{4.97}$$

Thus, the curves corresponding to the TE_{01} and TM_{01} modes[†] cross the horizontal axis at $u = 2.4048$. Similarly the curves corresponding to the TE_{02} and TM_{02} modes cross at $u = 5.5021$ (see Fig. 4.7).

For $l \neq 0$, the calculations become more involved, and one has to deal with the complete transcendental equation [Eq. (4.92)]. However, with some algebra and the use of recurrence relations, Eq. (4.92) can be transformed to the following[‡]:

$$(\mathscr{E}J^- - K^-)(J^+ - K^+) + (\mathscr{E}J^+ - K^+)(J^- - K^-) = 0 \tag{4.98}$$

---

*This follows from the relation

$$\lim_{w \to 0} \frac{w K_0(w)}{K_0'(w)} = \lim_{w \to 0} -\frac{w K_0(w)}{K_1(w)} = 0$$

[†]Here the first subscript 0 refers to $l = 0$ and the second subscript 1 refers to the fact that we are considering the first mode.

[‡]The details have been worked out by Marcuse (1972); the method was originally developed by Schlesinger *et al.* (1960).

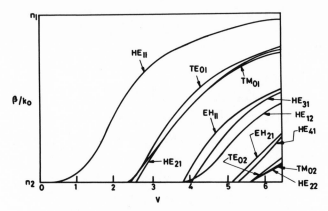

Fig. 4.7. Rigorous calculation of β/k_0 as a function of v for a practical fiber (after Maurer, 1974; reprinted by permission).

where

$$J^+ = \frac{1}{u}\frac{J_{l+1}(u)}{J_l(u)} \tag{4.99}$$

$$J^- = \frac{1}{u}\frac{J_{l-1}(u)}{J_l(u)} \tag{4.100}$$

$$K^+ = \frac{1}{w}\frac{K_{l+1}(w)}{K_l(w)} \tag{4.101}$$

$$K^- = \frac{1}{w}\frac{K_{l-1}(w)}{K_l(w)} \tag{4.102}$$

and

$$\mathscr{E} = \frac{n_1^2}{n_2^2}$$

For $w \to 0$, using the limiting forms of $K_0(w)$ and $K_1(w)$ [see Eqs. (4.20) and (4.21)], one obtains, for $l = 1$

$$J_1(u) = 0 \tag{4.103}$$

The corresponding propagation constants are plotted as HE_{11}, HE_{12}, etc., which intersect the horizontal axis at the zeros of $J_1(u)$, at $u = 0, 3.8317$, etc. Notice that the HE_{11} mode has no cutoff frequency and that the present cutoff conditions agree with the ones obtained by Gloge's approximate theory.

For $l = 2, 3, 4, \ldots$ and $u \to 0$ one obtains *either*

$$(\mathscr{E} + 1)J_{l-1}(u) - \frac{u}{(l-1)} \; J_l(u) = 0 \qquad (4.104)$$

or

$$J_l(u) = 0 \qquad (4.105)$$

where the solution

$$u = 0$$

should be excluded.

The modes whose cutoff frequencies are determined by the equation

$$J_l(u) = 0 \qquad l = 2, 3, \ldots \qquad (4.106)$$

are designated as EH_{lm} modes. For example, the zeros of $J_2(x)$ occur at $x = 5.1356, 8.4172$, etc.; thus the propagation characteristics corresponding to EH_{21} and EH_{22} modes intersect the horizontal axis at 5.1356 and 8.4172, respectively (see Fig. 4.7).

On the other hand, the modes whose cutoff frequencies are determined by the equation

$$(\mathscr{E} + 1)J_{m-1}(u) = \frac{u}{(m-1)} \; J_m(u) \qquad (4.107)$$

are designated as HE_{mn} modes.

The group velocities of various modes and their relation to pulse dispersion will be discussed in Chapter 7.

CHAPTER 5

Inhomogeneous Circular Waveguides

This chapter is devoted to the study of guided propagation of electromagnetic waves through a medium characterized by the following dielectric constant variation

$$K = K_0 - K_2(x^2 + y^2) \tag{5.1}$$

$$= K_0 - K_2 r^2 \tag{5.2}$$

where $r^2 = x^2 + y^2$. Such a medium is commonly referred to as a square-law medium. The fabrication of optical waveguides, in which transmission takes place through square-law media, has been discussed in Appendix B. Such waveguides in the form of fibers present exciting possibilities as transmitting channels for optical communications. The gas lenses are also characterized by a similar refractive index variation with K_2 depending on z. The main advantage of a square-law medium is that it introduces extremely small dispersion over long distances in any pulse traveling through it (see Chapter 7). The aberrations in images formed by such media are also small (see Chapter 9). In the present chapter we discuss the propagation of electromagnetic waves in such media on the basis of the scalar wave equation. The modifications introduced by the use of vector equations will be discussed in Chapter 6.

In Chapters 2 and 3 we have shown that in an inhomogeneous medium the scalar wave equation is valid when the variation of the dielectric constant is small in distances of the order of wavelength (which is true for practical optical waveguides). Thus the modes of propagation in the medium can be obtained from the scalar wave equation

$$\nabla^2\psi + (K\omega^2/c^2)\psi = 0 \tag{5.3}$$

where ψ is the amplitude of the electric vector or the magnetic vector.

As in the previous chapter, the dielectric constant variation given by Eqs. (5.1) and (5.2) cannot extend to infinity and there has to be a boundary (finite values of x and y) beyond which Eq. (5.4) will not be valid.* However, if the beam dimensions are such that a major portion of the beam is confined to a region where Eq. (5.1) is valid then we are justified in using Eq. (5.1) for all values of x and y; in other words,† the width of the beam is much less than the core diameter a. This is known as the infinite medium approximation. The effect of the presence of the core–cladding interface will be briefly discussed in Chapter 6.

5.1. Modal Analysis

For the sake of generality we will assume the spatial dependence of the dielectric constant to be of the form

$$K = K_0 - K_2 x^2 - K_3 y^2 \tag{5.5}$$

when $K_2 = K_3$, we obtain Eq. (5.1). If we substitute the above expression for K in Eq. (5.3) we obtain

$$\nabla^2\psi + (\omega^2/c^2)(K_0 - K_2 x^2 - K_3 y^2)\psi = 0 \tag{5.6}$$

We proceed to solve Eq. (5.6) by the method of separation of variables; thus

$$\psi = X(x)Y(y)Z(z) \tag{5.7}$$

On substituting for ψ from Eq. (5.7) in Eq. (5.6), we get

$$\left(\frac{1}{X}\frac{d^2X}{dx^2} - \frac{\omega^2}{c^2}K_2 x^2\right) + \left(\frac{1}{Y}\frac{d^2Y}{dy^2} - \frac{\omega^2}{c^2}K_3 y^2\right) + \left(\frac{1}{Z}\frac{d^2Z}{dz^2} + k^2\right) = 0 \tag{5.8}$$

---

*Indeed in an actual fiber the dielectric constant variation is of the form

$$K = K_0 - K_2 r^2 \qquad \text{for } r < a$$
$$= K_1 \qquad \text{for } r > a \tag{5.4}$$

†In the approximation of geometrical optics this condition implies that the rays do not impinge on the core–cladding interface ($r = a$).

where

$$k^2 = (\omega^2/c^2)K_0 \tag{5.9}$$

Thus we may write*

$$\frac{1}{X}\frac{d^2X}{dx^2} - \frac{\omega^2}{c^2}K_2x^2 = -\gamma_1^2 \tag{5.10a}$$

$$\frac{1}{Y}\frac{d^2Y}{dy^2} - \frac{\omega^2}{c^2}K_3y^2 = -\gamma_2^2 \tag{5.10b}$$

and

$$\frac{1}{Z}\frac{d^2Z}{dz^2} + k^2 - \gamma_1^2 - \gamma_2^2 = 0 \tag{5.10c}$$

Therefore

$$\frac{d^2X}{d\xi^2} + (\lambda - \xi^2)X(\xi) = 0 \tag{5.11a}$$

$$\frac{d^2Y}{d\eta^2} + (\mu - \eta^2)Y(\eta) = 0 \tag{5.11b}$$

and

$$\frac{d^2Z}{dz^2} + \beta^2 Z(z) = 0 \tag{5.11c}$$

where

$$\xi = \left(\frac{\omega^2 K_2}{c^2}\right)^{1/4}x = \frac{x}{x_0}$$

$$x_0 = \left(\frac{c^2}{\omega^2 K_2}\right)^{1/4} \tag{5.12a}$$

$$\lambda = (\gamma_1 x_0)^2$$

$$\eta = \left(\frac{\omega^2 K_3}{c^2}\right)^{1/4}y = \frac{y}{y_0}$$

$$y_0 = \left(\frac{c^2}{\omega^2 K_3}\right)^{1/4} \tag{5.12b}$$

$$\mu = (\gamma_2 y_0)^2$$

$$\beta^2 = k^2 - \gamma_1^2 - \gamma_2^2 \tag{5.13}$$

*If we assume the RHS of Eqs. (5.10a) and (5.10b) to be of the form $+\gamma_1^2$ and $+\gamma_2^2$, we obtain solutions which go to infinity $x \to \infty$ or as $x \to -\infty$. This can easily be seen if we put $K_2 = 0$; Eq. (5.10a) and (5.10b) lead to an exponential dependence of the field on x and y, which is certainly not admissible for a homogeneous medium.

Both Eqs. (5.11a) and (5.11b) are of the same form as the Schrödinger equation for a linear harmonic oscillator; for the fields to be finite as $x \to \pm\infty$ we must have (see, for example, Schiff, 1968)

$$\lambda = (2n + 1)$$
$$\mu = (2m + 1) \tag{5.14}$$

where n, $m = 0, 1, 2, \ldots$. The corresponding field patterns are given by

$$X_n(\xi) = C_n H_n(\xi) \exp(-\tfrac{1}{2}\xi^2)$$
$$Y_m(\eta) = D_m H_m(\eta) \exp(-\tfrac{1}{2}\eta^2) \tag{5.15}$$

where H_n are the Hermite polynomials and C_n and D_m are arbitrary constants. If the functions X_n and Y_m are assumed to be normalized, i.e.,

$$\int_{-\infty}^{\infty} |X_n(x)|^2 \, dx = 1 \tag{5.16a}$$

and

$$\int_{-\infty}^{\infty} |Y_m(y)|^2 \, dy = 1 \tag{5.16b}$$

then the constants would be given by

$$C_n = \left(\frac{1}{x_0} \frac{1}{2^n n! \sqrt{\pi}}\right)^{1/2} \tag{5.17a}$$

and

$$D_m = \left(\frac{1}{y_0} \frac{1}{2^m m! \sqrt{\pi}}\right)^{1/2} \tag{5.17b}$$

where x_0 and y_0 have been defined by Eq. (5.12). Using Eq. (5.14) the allowed values of β obtained from Eq. (5.13) are

$$\beta_{nm} = \left[k^2 - \frac{(2n+1)}{x_0^2} - \frac{(2m+1)}{y_0^2}\right]^{1/2}$$

$$= \frac{\omega}{c} K_0^{1/2} \left[1 - (2n+1)\frac{c}{\omega K_0} K_2^{1/2} - (2m+1)\frac{c}{\omega K_0} K_3^{1/2}\right]^{1/2} \tag{5.18}$$

For a typical SELFOC fiber $K_0 \simeq 2.5$ and $(K_2/K_0) \simeq 0.2 \text{ mm}^{-2}$ (see Appendix B); in the optical region $\lambda \simeq 6 \times 10^{-4}$ mm, giving

$$\frac{c}{\omega K_0} K_2^{1/2} \simeq \frac{6 \times 10^{-4}}{2\pi (2.5)^{1/2}} (0.2)^{1/2} \simeq 3 \times 10^{-5}$$

Thus, for most practical systems $(c/\omega K_0)K_2^{1/2} \ll 1$ and we are justified in making a binomial expansion of the RHS of Eq. (5.18), i.e.,

$$\beta_{nm} \approx \frac{\omega}{c} K_0^{1/2}\left[1 - (n+\tfrac{1}{2})\frac{c}{\omega K_0}K_2^{1/2} - (m+\tfrac{1}{2})\frac{c}{\omega K_0}K_3^{1/2}\right] \quad (5.19)$$

$$\approx \frac{\omega}{c} K_0^{1/2} - (n+\tfrac{1}{2})\left(\frac{K_2}{K_0}\right)^{1/2} - (m+\tfrac{1}{2})\left(\frac{K_3}{K_0}\right)^{1/2} \quad (5.20)$$

If we evaluate the group velocity $[=1/(d\omega/d\beta)]$ using Eq. (5.20), it is seen to be independent of m and n. Thus if the binomial expansion which leads to Eq. (5.19) is valid, the group velocity is independent of the nature of the mode (values of m and n) and one should not expect any dispersion of the pulse. Indeed, the pulse dispersion in SELFOC fibers is extremely small (see Chapter 7). This is one of the most important advantages of SELFOC fibers.

5.1.1. The Kernel

The general solution of the scalar wave equation [Eq. (5.5)] can be written in the form

$$\varphi(x, y, z) = \sum_{m,n} A_{nm}X_n(x)Y_m(y)\exp[-i\beta_{nm}z] \quad (5.21)$$

where the constants A_{nm} are to be determined from initial conditions; for example, if the field pattern is known at $z = 0$ then

$$A_{nm} = \int_{-\infty}^{\infty}\int_{-\infty}^{\infty} X_n(x)Y_m(y)\varphi(x, y, 0)\,dx\,dy \quad (5.22)$$

where use has been made of the fact that the functions $X_n(x)$ and $Y_m(y)$ form an orthonormal set, i.e.,

$$\int_{-\infty}^{\infty} X_m(x)X_n(x)\,dx = \delta_{nm} \quad (5.23a)$$

and

$$\int_{-\infty}^{\infty} Y_m(y)Y_n(y)\,dy = \delta_{nm} \quad (5.23b)$$

If we substitute for A_{nm} from Eq. (5.22) in Eq. (5.21) we obtain

$$\varphi(x, y, z) = \iint K(x, y, x', y', z)\varphi(x', y', 0)\,dx'\,dy' \quad (5.24)$$

where

$$K(x, y, x', y', z) = \sum_{n=0,1,2,\dots} \sum_{m=0,1,2,\dots} X_n(x)X_n(x')$$

$$\times Y_m(y)Y_m(y')\exp(-i\beta_{nm}z) \qquad (5.25)$$

is known as the kernel. If we use the approximate form of β_{nm}, as given by Eq. (5.20), the kernel can be written as a product of two sums:

$$K(x, y, x', y', z) = \exp\left\{-i\left[\frac{\omega}{c}K_0^{1/2} - \frac{1}{2}\left(\frac{K_2}{K_0}\right)^{1/2} - \frac{1}{2}\left(\frac{K_3}{K_0}\right)^{1/2}\right]z\right\}$$

$$\times F_1(x, x', z)F_2(y, y', z) \qquad (5.26)$$

where

$$F_1(x, x', z) = \sum_{n=1,1,2,\dots} X_n(x)X_n(x')\exp\left[in\left(\frac{K_2}{K_0}\right)^{1/2}z\right] \qquad (5.27a)$$

and

$$F_2(y, y', z) = \sum_{m=0,1,2,\dots} Y_m(y)Y_m(y')\exp\left[im\left(\frac{K_3}{K_0}\right)^{1/2}z\right] \qquad (5.27b)$$

If one substitutes the Hermite–Gauss functions for X_n and Y_m, the resulting series can be summed by using Mehler's formula [see Eq. (3.24)]; the final results are

$$F_1(x, x', z) = \frac{1}{x_0\sqrt{\pi}(1-\zeta_2^2)^{1/2}}\exp\left[\frac{2\xi\xi'\zeta_2}{1-\zeta_2^2} - \frac{(\xi^2+\xi'^2)(1+\zeta_2^2)}{2(1-\zeta_2^2)}\right] \qquad (5.28a)$$

and

$$F_2(y, y', z) = \frac{1}{y_0\sqrt{\pi}(1-\zeta_3^2)^{1/2}}\exp\left[\frac{2\eta\eta'\zeta_3}{1-\zeta_3^2} - \frac{(\eta^2+\eta'^2)(1+\zeta_3^2)}{2(1-\zeta_3^2)}\right] \qquad (5.28b)$$

where

$$\zeta_2 = \exp[i(K_2/K_0)^{1/2}z] \qquad (5.29)$$

and

$$\zeta_3 = \exp[i(K_3/K_0)^{1/2}z] \qquad (5.29)$$

and ξ and η have been defined through Eq. (5.12). Since the form of the kernel is known [see Eq. (5.26)] one can [using Eq. (5.24)] calculate the intensity distribution of the beam as it propagates through the medium. In what follows we assume $K_2 = K_3$ (which is indeed the case for SELFOC fibers). Thus

$$x_0 = y_0 = \alpha \quad \text{(say)} \tag{5.30a}$$

and

$$\zeta_2 = \zeta_3 = \exp[i(K_2/K_0)^{1/2}z] = \exp(i\delta z) \quad \text{(say)} \tag{5.30b}$$

where

$$\delta = \left(\frac{K_2}{K_0}\right)^{1/2} \tag{5.30c}$$

Further

$$1 - \zeta_2^2 = 1 - \exp(2i\delta z) = -2i \exp(i\delta z) \sin \delta z$$

and

$$1 + \zeta_2^2 = 1 + \exp(2i\delta z) = 2 \exp(i\delta z) \cos \delta z$$

On substituting for F_1 and F_2 from Eqs. (5.28a) and (5.28b) in Eq. (5.26) and carrying out some simplification, we finally obtain

$$K(x, y, x', y', z) = \frac{1}{\alpha^2} \frac{\exp[-i(\omega/c)K_0^{1/2}z - (i/2)(\xi^2 + \eta^2) \cot \delta z]}{2\pi i \sin \delta z}$$

$$\times \exp\left[\frac{i}{\sin \delta z}(\xi\xi' + \eta\eta') - \frac{i}{2} \cot \delta z(\xi'^2 + \eta'^2)\right] \tag{5.31}$$

5.1.2. Propagation of a Gaussian Beam Incident Normally at an Off-Axis Point

Let us consider a simple application of the above result, namely, the normal incidence of a Gaussian beam; the center of the beam is an off-axis point:

$$\varphi(x, y, z = 0) = A \exp\left[-\frac{(x-a)^2 + y^2}{2\alpha^2}\right]$$

$$= A \exp\left[-\frac{(\xi - a_0)^2 + \eta^2}{2}\right] \tag{5.32}$$

where $a_0 = a/\alpha$, and the x axis is chosen along the displacement (since the system is cylindrically symmetric the x and y directions are arbitrary). For the sake of mathematical simplicity, the beamwidth has been assumed to be the same as $\alpha(=x_0 = y_0)$ defined by Eq. (5.30a); the analysis can be generalized for an arbitrary beamwidth. On substituting the expressions for

$K(x, y, x', y', z)$ and $\varphi(x, y, z = 0)$ from Eqs. (5.31) and (5.32) in Eq. (5.24) we obtain

$$\varphi(x, y, z) = -\frac{A \exp[-i(\omega/c)K_0^{1/2}z - (i/2)(\xi^2 + \eta^2) \cot \delta z]}{2\pi i \sin \delta z}$$

$$\times \int_{-\infty}^{\infty} \exp\left[-\frac{(\xi' - a_0)^2}{2} + \frac{i}{\sin \delta z}\xi\xi' - i\frac{\xi'^2}{2} \cot \delta z\right] d\xi'$$

$$\times \int_{-\infty}^{\infty} \exp\left(-\frac{\eta'^2}{2} + \frac{i}{\sin \delta z}\eta\eta' - i\frac{\eta'^2}{2} \cot \delta z\right) d\eta' \qquad (5.33)$$

The integrations can be carried out by using the method discussed in Chapter 3. The final result is

$$\varphi(x, y, z) = A \exp\left[-\frac{(x - a \cos \delta z)^2 + y^2}{2\alpha^2}\right]$$

$$\times \exp\left[\frac{i}{2\alpha^2}(2xa \sin \delta z - \frac{a^2}{2} \sin 2\delta z)\right]$$

$$\times \exp\{-i[(\omega/c)K_0^{1/2} - \delta]z\} \qquad (5.34)$$

Thus the center of the field distribution follows the path

$$x = a \cos \delta z$$
$$y = 0 \qquad (5.35)$$

which describes the path of a ray. The intensity distribution remains Gaussian and the beam follows a sinusoidal path as it propagates through the fiber. This is also what is experimentally observed (see, for example, Kita and Uchida, 1969); we will consider this in more detail in Sec. 5.2.

We can also obtain closed form solutions for the case when the initial intensity distribution is of the form

$$\varphi(x, y, z = 0) = \varphi_0 \exp\left(-\frac{r^2}{2w_0^2}\right)$$

$$= \varphi_0 \exp\left(-\frac{x^2 + y^2}{2w_0^2}\right) \qquad (5.36)$$

$$= \varphi_0 \exp\left(-\frac{\xi^2 + \eta^2}{2\sigma^2}\right)$$

where $\sigma = w_0/\alpha$ with w_0 representing the initial beamwidth. The above equation may be compared with Eq. (3.26). The analysis can be carried out in an exactly similar manner. The final result is [cf. Eq. (3.30)].

$$\varphi(x, y, z) = \frac{\sqrt{2}\varphi_0\sigma^2 \exp[i\chi(z)]}{[\sigma^4+1+(\sigma^4-1)\cos 2\delta z]^{1/2}}$$

$$\times \exp\left\{-\frac{(\xi^2+\eta^2)\sigma^2}{[\sigma^4+1+(\sigma^4-1)\cos 2\delta z]}\right\} \quad (5.37)$$

where

$$\chi(z) = -\frac{\omega}{c}K_0^{1/2}z + \tan^{-1}\left(\frac{\tan \delta z}{\sigma^2}\right)$$

$$+\frac{(\xi^2+\eta^2)}{2}\frac{(\sigma^4-1)\sin 2\delta z}{[\sigma^4+1+(\sigma^4-1)\cos 2\delta z]} \quad (5.37a)$$

Thus

$$|\varphi(x, y, z)|^2 = |\varphi_0|^2 \frac{w_0^2}{w^2(z)} \exp\left[-\frac{x^2+y^2}{w^2(z)}\right] \quad (5.38)$$

where

$$w^2(z) = \frac{w_0^2}{2\sigma^4}[\sigma^4+1+(\sigma^4-1)\cos 2\delta z] \quad (5.39)$$

Thus the intensity distribution remains Gaussian. When $\sigma = 1$ (i.e., $w_0 = \alpha$) then $w(z) = w_0 = \alpha$ for all values of z, and the beam propagates as the fundamental mode.

5.1.3. Path of Rays: A General Treatment*

To find the path of a ray, we assume a Gaussian beam (of halfwidth w_0) incident on the plane $z = 0$, at a distance a from the axis and in any general direction. Consider two coordinate frames, one (x, y, z) fixed in the medium, z being along the axis of the fiber; and the other (x', y', z') fixed on the incident Gaussian beam, z' being along the initial direction of propagation. The relation between the two frames can be specified by the Euler angles† (θ, ϕ, ψ). Then

$$\begin{pmatrix} x' \\ y' \\ z' \end{pmatrix} = \begin{pmatrix} b_{11} & b_{12} & b_{13} \\ b_{21} & b_{22} & b_{23} \\ b_{31} & b_{32} & b_{33} \end{pmatrix} \begin{pmatrix} x-a \\ y \\ z \end{pmatrix} \quad (5.40)$$

*After Ghatak and Thyagarajan (1975).
†The notation is consistent with that of Goldstein (1950), except for the fact that there has been an anticyclic rotation of the three axes (see Fig. 5.1).

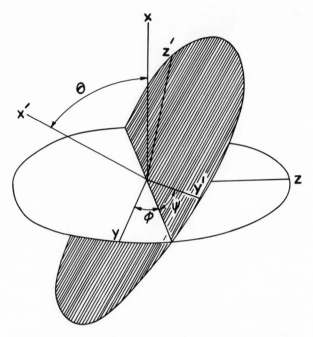

Fig. 5.1. Definition of θ, φ, and ψ (after Ghatak and Thyagarajan, 1975; reprinted by permission).

where

$$b_{11} = \cos \theta \qquad b_{12} = \sin \theta \sin \phi. \qquad (5.41)$$

$$b_{13} = -\sin \theta \cos \phi \qquad b_{21} = \sin \psi \sin \theta \qquad (5.42)$$

$$b_{22} = \cos\psi \cos \phi - \sin \psi \cos \theta \sin \phi \qquad (5.43)$$

$$b_{23} = \cos \psi \sin \phi + \sin \psi \cos \theta \cos \phi \qquad (5.44)$$

$$b_{31} = \cos \psi \sin \theta \qquad (5.45)$$

$$b_{32} = -\sin \psi \cos \phi - \cos \psi \cos \theta \sin \phi \qquad (5.46)$$

and

$$b_{33} = -\sin \psi \sin \phi + \cos \psi \cos \theta \cos \phi \qquad (5.47)$$

The displacement, a, is chosen along the x axis. This, however, does not affect the generality of the problem. The incident Gaussian beam can be written as

$$\Psi(x', y', z') = \Psi_0 \exp\left(-\frac{x'^2 + y'^2}{2w_0^2}\right) \exp(-ikz') \qquad (5.48)$$

Thus

$$\Psi(x, y, z = 0) = \Psi_0 \exp\left(-\frac{p}{2\sigma^2}\xi^2 - \frac{q}{2\sigma^2}\eta^2 - \frac{s}{\sigma^2}\xi\eta + ik\alpha g\xi + ik\alpha h\eta\right)$$

$$\times \exp\left(-\frac{p}{2w_0^2}a^2 + ikb_{31}a\right) \qquad (5.49)$$

where

$$p = b_{11}^2 + b_{21}^2 \qquad\qquad q = b_{12}^2 + b_{22}^2$$

$$s = b_{11}b_{12} + b_{21}b_{22} \qquad g = -b_{31} - \frac{ipa}{kw_0^2} \qquad (5.50)$$

$$h = -b_{32} - \frac{isa}{kw_0^2} \qquad\qquad \sigma = \frac{w_0}{\alpha}$$

On substituting the expression for Ψ from Eq. (5.49) in Eq. (5.24) and using Eq. (5.31) we obtain, after integration, the amplitude at any point (x, y, z) as

$$\Psi(x, y, z) = -\Psi_0 \frac{\exp[-i(\omega/c)K_0^{1/2}z]}{i \sin \delta z} \exp\left(-\frac{p}{2w_0^2}a^2 + ikb_{31}a\right)$$

$$\times \frac{1}{[(q/\sigma^2) + i \cot \delta z]^{1/2}} \frac{\exp[-(i/2)(\xi^2 + \eta^2) \cot \delta z]}{\left\{\dfrac{p}{\sigma^2} + i \cot \delta z - \dfrac{s^2}{\sigma^4[(q/\sigma^2) + i \cot \delta z]}\right\}^{1/2}}$$

$$\times \exp\left\{-\frac{[kh\alpha + (\eta/\sin \delta z)]^2}{2[(q/\sigma^2) + i \cot \delta z]}\right\}$$

$$\times \exp\left\{-\frac{\left\{k\alpha g + \dfrac{\xi}{\sin \delta z} - \dfrac{s[kh\alpha + (\eta/\sin \delta z)]}{\sigma^2[(q/\sigma^2) + i \cot \delta z]}\right\}^2}{2\left\{\dfrac{p}{\sigma^2} + i \cot \delta z - \dfrac{s^2}{\sigma^4[(q/\sigma^2) + i \cot \delta z]}\right\}}\right\} \qquad (5.51)$$

We consider some specific examples.

A Beam Displaced from the Axis and Incident Parallel to It. For a beam of halfwidth α displaced from the axis and incident parallel to it, we have $\theta = 0$, $\phi = 0$, $\psi = 0$, $p = 1$, $q = 1$, $s = 0$, $g = -ia/k\alpha^2$, and $h = 0$. The expression for Ψ reduces to

$$\Psi = \Psi_0 \exp\left[-\frac{(x - a \cos \delta z)^2 + y^2}{2\alpha^2}\right] \exp\left[-i\left(\frac{\omega}{c}K_0^{1/2} - \delta\right)z\right]$$

$$\times \exp\left[\frac{i}{2\alpha^2}\left(2xa \sin \delta z - \frac{a^2}{2} \sin 2\delta z\right)\right]$$

This result is the same as the one given by Eq. (5.34).

An Incident On-Axis Gaussian Beam with Arbitrary Halfwidth. We consider a Gaussian beam of width w_0 incident along the z axis. Then $\theta = 0$, $\phi = 0$, $\psi = 0$, and $a = 0$; and Eq. (5.51) reduces to Eq. (5.37).

Helical Rays. To study the propagation of helical rays, a special kind of skew rays that travel always at the same distance from the axis, we consider the set of rays specified by $\theta = 0$. These are a special set of rays displaced from the axis and incident at different angles from the z axis along the y–z plane. Of these rays the helical rays are formed under a special condition (on the angle and the displacement from the axis) that will now be derived.

For $\theta = 0$ and $w_0 = \alpha$ we have $p = 1$, $q = \cos^2 \chi$, $s = 0$, $g = -ia/k\alpha^2$, and $h = \sin \chi$, where $\chi = \phi + \psi$ is the angle between the incident ray and the z axis. Then Eq. (5.51) reduces to

$$
\Psi(x, y, z) = \frac{\Psi_0 \exp\left[-i\left(\frac{\omega}{c}K_0^{1/2} - \frac{\delta}{2}\right)z\right]}{(\cos^2 \delta z + \sin^2 \delta z \cos^4 \chi)^{1/4}}
$$

$$
\times \exp\left[\frac{i}{2}\tan^{-1}(\cos^2 \chi \tan \delta z)\right] \exp\left(-\frac{i}{2}\eta^2 \cot \delta z\right)
$$

$$
\times \exp\left\{-\frac{(\eta + k\alpha \sin \chi \sin \delta z)^2}{2[\sin^2 \delta z + (\cos^2 \delta z/\cos^2 \chi)]} - \frac{1}{2}\left(\xi - \frac{a}{\alpha}\cos \delta z\right)^2\right\}
$$

$$
\times \exp\left[-\frac{i}{2}\frac{(\eta + k\alpha \sin \chi \sin \delta z)^2}{(\cos^2 \chi \sin^2 \delta z + \cos^2 \delta z)}\cot \delta z\right]
$$

$$
\times \exp\left[\frac{i}{2}\left(\frac{2a\xi}{\alpha}\sin \delta z - \frac{a^2}{2\alpha^2}\sin 2\delta z\right)\right] \tag{5.52}
$$

Hence the ray path is

$$
x = a \cos \delta z
$$
$$
y = -(\sin \chi/\delta) \sin \delta z \tag{5.53}
$$

The distance r of any point on the ray from the axis is given by

$$
r^2 = x^2 + y^2
$$
$$
= a^2 \cos^2 \delta z + (\sin^2 \chi/\delta^2) \sin^2 \delta z \tag{5.54}
$$

which, for helical rays, has to be independent of z. Hence r will be independent of z only when

$$
\sin \chi = a\delta \tag{5.55}
$$

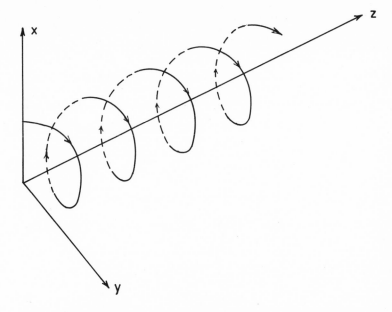

Fig. 5.2. The propagation of a special kind of skew rays, namely, the helical rays, through a square-law medium. These rays always travel at a constant distance from the z axis and thus trace out a helix. Such rays are formed when Eq. (5.55) is satisfied.

This represents the special condition under which a general skew ray will be a helical ray in a square-law medium (see Fig. 5.2).

5.1.4. Three-Dimensional Intensity Pattern Near Focus

We can use the formulation developed above to study the three-dimensional intensity distribution near the focus. To be able to compare it with the three-dimensional intensity pattern produced by a lens (Born and Wolf, 1970, p. 435), we consider a truncated plane wave of circular cross section incident parallel to the axis, with its center coinciding with $x = 0$, $y = 0$. We change over to cylindrical coordinates in Eq. (5.31) by putting

$$\xi = (r/\alpha) \cos \theta$$
$$\eta = (r/\alpha) \sin \theta$$

$$(5.56)$$

θ being the azimuthal coordinate. Hence we get

$$\Psi(r, \theta, z) = -\frac{\exp[-i(\omega/c)K_0^{1/2}z]}{2\pi i \sin \delta z} \exp\left(-\frac{i}{2}\frac{r^2}{\alpha^2}\cot \delta z\right)$$

$$\times \int_0^\infty dr' \int_0^{2\pi} d\theta' \, \Psi(r', \theta', z = 0)$$

$$\times \exp\left[\frac{irr'}{\alpha^2 \sin \delta z}\cos(\theta - \theta') - i\frac{r'^2}{2\alpha^2}\cot \delta z\right]\frac{r'}{\alpha^2} \qquad (5.57)$$

Now,

$$\Psi(r', \theta', z = 0) = \Psi_0 \qquad \text{for } r' \leq a$$
$$= 0 \qquad \text{for } r' > a \qquad (5.58)$$

where a is the radius of cross section of the beam. To reduce the equation to a familiar form, we substitute

$$\rho^2 = \xi'\frac{\alpha^2}{a^2}$$

$$v = \frac{a}{\alpha^2}\frac{r}{\sin \delta z}$$

$$u = \frac{a^2}{\alpha^2}\cot \delta z$$

then Eq. (5.57) reduces to

$$\Psi(r, z) = \Psi_0 a^2 \frac{\exp[-i(\omega/c)K_0^{1/2}z]}{2i\alpha^2 \sin \delta z} \exp\left(-i\frac{r^2}{2\alpha^2}\cot \delta z\right)$$

$$\times 2\int_0^1 \exp\left(-\frac{i}{2}u\rho^2\right)J_0(v\rho)\rho \, d\rho \qquad (5.59)$$

For points around the focus, i.e., $\delta z = (2n + 1)\pi/2$ $(n = 0, 1, 2, \ldots)$, this intensity distribution is exactly similar to the one produced by a simple lens at its focus (Born and Wolf, 1970). Since the medium is a continuously focusing one, the distribution of intensity as z increases to values near the planes given by $\delta z = n\pi$ $(n = 1, 2, \ldots)$ resembles completely the incident intensity distribution.

From Eq. (5.59) one can easily deduce the intensity distribution at the focal planes, given by $\delta z = (2n + 1)\pi/2$:

$$I(r) = |\Psi(r)|^2$$

$$= I_0 \left|\frac{2J_1[(a/\alpha^2)r]}{(a/\alpha^2)r}\right|^2 \qquad (5.60)$$

where $I_0 = |(\Psi_0 a^2/2\alpha^2)|^2$ is the intensity at the focus. Equation (5.60) is just the Airy pattern.

Similarly, the intensity distribution along the axis can be determined to be

$$I(r = 0, z) = I_0 \left| \frac{\sin(u/4)}{(u/4)} \right|^2 + 16 I_0 \frac{\alpha^4}{a^4} \sin^2 \frac{u}{4} \qquad (5.61)$$

This pattern differs from that produced by a simple lens in the presence of the second term in Eq. (5.61), the difference depends on the parameter a/α, the ratio of the incident beam radius to that of the fundamental beam mode width of the medium.

5.1.5. Effect of Absorption

In an absorbing medium K_0 is complex and Eq. (5.5) may be written as

$$K = K_{0r} - i K_{0i} - K_2(x^2 + y^2) \qquad (5.62)$$

We assume K_2 to be real. The modal analysis approach can be extended to this case and one finds that there is no change in the field patterns corresponding to various modes; however, the propagation constants in this case are given by

$$\beta_{mn} = \frac{\omega}{c}(K_{0r} - iK_{0i})^{1/2}$$

$$\times \left[1 - (2n+1) \frac{cK_2^{1/2}}{\omega(K_{0r} - iK_{0i})} - (2m+1) \frac{cK_2^{1/2}}{\omega(K_{0r} - iK_{0i})} \right]^{1/2}$$

or

$$\beta_{mn} \simeq \frac{\omega}{c}(K_{0r})^{1/2} - i\frac{\omega}{c} \frac{K_{0i}}{2(K_{0r})^{1/2}}$$

$$- (n + \tfrac{1}{2}) \left[\left(\frac{K_2}{K_{0r}} \right)^{1/2} + \frac{i}{2} \left(\frac{K_2}{K_{0r}} \right)^{1/2} \frac{K_{0i}}{K_{0r}} \right] \qquad (5.63)$$

$$- (m + \tfrac{1}{2}) \left[\left(\frac{K_2}{K_{0r}} \right)^{1/2} + \frac{i}{2} \left(\frac{K_2}{K_{0r}} \right)^{1/2} \frac{K_{0i}}{K_{0r}} \right]$$

where K_{0i} is assumed to be small in comparison to K_{0r}. Clearly, the attenuation corresponding to the fundamental mode ($n = m = 0$) will be minimum and therefore the fundamental mode will dominate at large z

values. The kernel will be of the form

$$K(\xi, \eta, \xi', \eta', z) = \frac{1}{\pi\alpha^2} \exp\left[-\left(\frac{\omega}{c}\frac{K_{0i}}{2K_{0r}^{1/2}} + \frac{1}{2}\left(\frac{K_2}{K_{0r}}\right)^{1/2}\frac{K_{0i}}{K_{0r}}\right)z\right]$$
$$\times \exp\left\{-\tfrac{1}{2}(\xi^2 + \xi'^2 + \eta^2 + \eta'^2)\right.$$
$$\left. - i\left[\frac{\omega}{c}K_{0r}^{1/2} - \left(\frac{K_2}{K_{0r}}\right)^{1/2}\right]z\right\}$$
$$\times \left[\sum_n \frac{\zeta_c^n}{2^n n!} H_n(\xi) H_n(\xi')\right]$$
$$\times \left[\sum_m \frac{\zeta_c^m}{2^m m!} H_m(\eta) H_m(\eta')\right] \tag{5.64}$$

where ζ_c is defined by the following equation

$$\zeta_c = \exp\left[i\left(\frac{K_2}{K_{0r}}\right)^{1/2} z - \frac{1}{2}\left(\frac{K_2}{K_{0r}}\right)^{1/2}\frac{K_{0i}}{K_{0r}}z\right] \tag{5.65}$$

Thus, using Mehler's formula [see Eq. (3.124)], we obtain

$$K(\xi, \eta, \xi', \eta', z) = \frac{1}{\pi\alpha^2(1-\zeta_c^2)}$$
$$\times \exp\left[-\left(\frac{\omega}{c}\frac{K_{0i}}{K_{0r}^{1/2}} + \frac{1}{2}\left(\frac{K_2}{K_{0r}}\right)^{1/2}\frac{K_{0i}}{K_{0r}}\right)z\right]$$
$$\times \exp\left[-\tfrac{1}{2}(\xi^2 + \xi'^2 + \eta^2 + \eta'^2) - i\gamma z\right]$$
$$\times \exp\left[\frac{2(\xi\xi' + \eta\eta')\zeta_c - (\xi^2 + \xi'^2 + \eta^2 + \eta'^2)\zeta_c^2}{1-\zeta_c^2}\right] \tag{5.66}$$

where $\gamma = (\omega/c)(K_{0r})^{1/2} - (K_2/K_{0r})^{1/2}$. For an incident beam having a field distribution of the form

$$\varphi(x, y, z=0) = \varphi_0 \exp\left(-\frac{\xi^2 + \eta^2}{2\sigma^2}\right)$$

the intensity distribution (as it propagates through the fiber) will remain Gaussian

$$I = \frac{I_0}{w^2(z)} \frac{4w_0^2\sigma^2 \exp\left[-\left(\frac{\omega}{c}\frac{K_{0i}}{K_{0r}^{1/2}} + \left(\frac{K_2}{K_{0r}}\right)^{1/2}\frac{K_{0i}}{K_{0r}}\right)z\right]}{[(\sigma^2+1)^2 - C^2(\sigma^2-1)^2]} \exp\left(-\frac{x^2+y^2}{w^2(z)}\right)$$

where

$$w^2(z) = \frac{w_0^2}{\sigma^2}\left[\frac{(\sigma^2+1)^2 + C^2(\sigma^2-1)^2 + 2C(\sigma^4-1)\cos 2\delta z}{(\sigma^2+1)^2 - C^2(\sigma^2-1)^2}\right]$$

with

$$C = \exp\left[-\left(\frac{K_2}{K_{0r}}\right)^{1/2}\frac{K_{0i}}{K_{0r}}z\right]$$

$$\delta = \left(\frac{K_2}{K_{0r}}\right)^{1/2}$$

and

$$w_0 = \sigma\alpha$$

This equation reduces to Eq. (5.39) in the absence of absorption, namely as $C \to 1$. It is also evident from Eq. (5.67) that in the limit $z \to \infty$, $C \to 0$ and the beamwidth tends to a value equal to the fundamental beamwidth of the medium.

5.2. The Parabolic Equation Approach

In general, it is quite cumbersome to use the modal analysis when K_0 and K_2 are dependent on z (which is indeed true for gas lenses and conical SELFOC fibers). Recently an alternative method has been suggested (Sodha *et al.*, 1971a,b) which gives an approximate analytical solution of

$$\nabla^2\psi + (\omega^2/c^2)(K_0 - K_2 r^2)\psi = 0 \qquad (5.68)$$

this solution is in agreement with the result obtained in Sec. 5.1.2 by using the modal analysis for the case $K_2(z) = $ constant. However, this method is based on the assumption of the field having azimuthal symmetry and also uses the WKB approximation, in which one assumes

$$\left|\frac{\partial^2 A}{\partial z^2}\right| \ll \left|k\frac{\partial A}{\partial z}\right| \qquad (5.69)$$

which is true whenever the focal length is much greater than the wavelength (slowly converging/diverging beam); A is defined in Eq. (5.70). The mathematical technique is similar to the one used by Akhmanov *et al.* (1966, 1968) in their investigation of self-focusing of laser beams in non-linear media.

We look for a solution of the scalar wave equation [Eq. (5.68)] of the form

$$\psi(r, z) = A(r, z)[K_0(0)/K_0(z)]^{1/4}\exp\left(-i\int k\,dz\right) \qquad (5.70)$$

where we have assumed ψ to have azimuthal symmetry and

$$k^2(z) = \frac{K_0(z)\omega^2}{c^2}$$ (5.71)

Substituting for ψ from Eq. (5.70) in Eq. (5.68), we obtain

$$\left(+2ik + \frac{1}{k}\frac{dk}{dz}\right)\frac{\partial A}{\partial z} = \frac{\partial^2 A}{\partial r^2} + \frac{1}{r}\frac{\partial A}{\partial r} - \left[k^2\frac{K_2(z)}{K_0(z)}\right]r^2 A$$ (5.72)

where use has been made of the relation [see Eq. (5.71)]:

$$\frac{2}{k}\frac{dk}{dz} = \frac{1}{K_0(z)}\frac{dK_0}{dz}$$ (5.73)

Further, in writing Eq. (5.72), the cylindrical system of coordinates has been used and terms proportional to $(\partial^2 A/\partial z^2)$, $(d/dz)[(1/K_0)(dK_0/dz)]$ and $[(1/K_0)(dK_0/dz)]^2$ have been neglected. This is justified for all practical systems corresponding to optical wavelengths. Equation (5.72) is often referred to as the parabolic equation. To solve Eq. (5.72) we introduce the eikonal S as

$$A(r, z) = A_0(r, z)\exp[-ikS(r, z)]$$ (5.74)

where A_0 and S are both real. If we substitute for A from Eq. (5.74) in Eq. (5.72) and equate the real and imaginary parts on both sides of the resulting equation, we obtain

$$2\frac{\partial S}{\partial z} + \frac{2S}{k}\frac{dk}{dz} + \left(\frac{\partial S}{\partial r}\right)^2 = -\frac{K_2(z)}{K_0(z)}r^2 + \frac{1}{k^2 A_0}\left[\frac{\partial^2 A_0}{\partial r^2} + \frac{1}{r}\frac{\partial A_0}{\partial r}\right]$$ (5.75)

and

$$\frac{\partial A_0^2}{\partial z} + \frac{\partial A_0^2}{\partial r}\frac{\partial S}{\partial r} + \left(\frac{\partial^2 S}{\partial r^2} + \frac{1}{r}\frac{\partial S}{\partial r}\right)A_0^2 = 0$$ (5.76)

where second-order terms like $(dk/dz)(\partial A_0/\partial z)$ have been neglected.

It is customary to talk of the two terms on the RHS of Eq. (5.75) as forces which determine the behavior of the eikonal, i.e., the convergence or the divergence of the beam (as discussed later). The first term is referred to as the inhomogeneous refraction force, while the second term is called the diffraction force. It may be pointed out that the word force has been used only in a colloquial sense and that these terms do not have the dimension MLT^{-2}.

Rigorous solution of Eqs. (5.75) and (5.76) is not possible and hence we look for approximate solutions. In the absence of inhomogeneity and diffraction, Eq. (5.75) reduces to

$$2\frac{\partial S}{\partial z} + \left(\frac{\partial S}{\partial r}\right)^2 = 0 \tag{5.77}$$

If we substitute a separable solution

$$S(r, z) = \mathscr{R}(r)\mathscr{L}(z)$$

in Eq. (5.77), we find that the variables indeed separate out as

$$-\frac{1}{\mathscr{L}^2}\frac{d\mathscr{L}}{dz} = \frac{1}{2\mathscr{R}(r)}\left(\frac{d\mathscr{R}}{dr}\right)^2 = C_1 \quad \text{(say)}$$

where C_1 is a constant. Thus

$$-\frac{d\mathscr{L}}{\mathscr{L}^2} = C_1\, dz$$

and

$$\frac{d\mathscr{R}}{[\mathscr{R}(r)]^{1/2}} = \pm(2C_1)^{1/2}\, dr$$

Simple integrations give us

$$S(r, z) = \frac{(\pm(C_1/2)^{1/2}r + C_3)^2}{(C_1 z + C_2)} \tag{5.78}$$

where C_3 and C_2 are constants of integration. With a suitable choice of origin, Eq. (5.78) can be put in the form

$$S(r, z) = \frac{r^2}{2(z + R)} \tag{5.79}$$

Equation (5.79) represents approximately a spherical wave; whose radius of curvature of the wavefront at $z = 0$ is R. This follows from the fact that for a spherical wave the phase factor is of the form $\exp(-ik\rho)$, where ρ is the distance measured from the center of the spherical wave front. Thus for Eqs. (5.70) and (5.74) to represent a spherical wave we must have*

$$S + z = \rho$$
$$= [r^2 + (R + z)^2]^{1/2} \tag{5.80}$$

---

*It should be remembered that we are, at the moment, looking for solutions in a homogeneous medium; as such k is independent of z.

where the center of the spherical wave front is assumed to be at $z = -R$. Clearly, the radius of curvature of the wave front at $z = 0$ is R. Further, for $r \ll R + z$ we can write

$$S + z = (R + z)\left[1 + \frac{r^2}{(R + z)^2}\right]^{1/2}$$

$$\approx (R + z) + \frac{r^2}{2(R + z)}$$

or

$$S \approx R + \frac{r^2}{2(R + z)} \tag{5.81}$$

which is of the same form* (within an additive constant factor to which S is always arbitrary) as given by Eq. (5.79).

Further, the radius of curvature of the wave front is given by

$$\frac{1}{R} = \frac{1}{r}\frac{\partial S}{\partial r} \tag{5.82}$$

(see Sodha et al., 1974), and therefore the eikonal given by Eq. (5.79) represents approximately a spherical wave, the radius of curvature of which, at $z = 0$, is R.

For a slightly converging/diverging wave, it is convenient to assume the following solution (which is similar to the one above)

$$S = (r^2/2)\beta(z) + \phi(z) \tag{5.83}$$

where $[\beta(z)]^{-1}$ represents the radius of curvature of the wave front. A number of solutions for S similar to that for a spherical wave may be assumed but Eq. (5.83) represents a mathematically convenient form. If S is given by Eq. (5.83) then Eq. (5.76) takes the form

$$\frac{\partial A_0^2}{\partial z} + r\left(\frac{1}{f}\frac{df}{dz}\right)\frac{\partial A_0^2}{\partial r} + 2\left(\frac{1}{f}\frac{df}{dz}\right)A_0^2 = 0 \tag{5.84}$$

where the function $f(z)$ is defined by the following equation

$$\frac{1}{f}\frac{df}{dz} = \beta(z) \tag{5.85}$$

*It is worthwhile to mention that if we do not neglect the term proportional to $\partial^2 A/\partial z^2$, then instead of Eq. (5.75) we would have obtained

$$\left(\frac{\partial S}{\partial z}\right)^2 + 2\left(\frac{\partial S}{\partial z}\right) + \left(\frac{\partial S}{\partial r}\right)^2 = 0$$

of which $S = [r^2 + (R + z)^2]^{1/2} - z$ is an exact solution.

The general solution of Eq. (5.84) is of the form (see Appendix C)

$$A_0^2 = \frac{E_0^2}{f^2} \mathcal{F}\left[\frac{r}{af(z)}\right] \tag{5.86}$$

where $\mathcal{F}$ is an arbitrary function of its argument.

It is instructive and useful to consider an alternative derivation for Eq. (5.86). Let a ray be represented by

$$r = \rho f(z) \tag{5.87}$$

where ρ represents the r coordinate of the ray at $z = 0$, i.e., $f = 1$ at $z = 0$. Hence the energy incident on a ring (perpendicular to the axis) formed by circles of radii ρ and $\rho + d\rho$ at $z = 0$ will be also incident on a ring formed by circles of radii r and $r + dr$ at z; we have of course assumed $f = 1$ at $z = 0$. Hence if $A_0^2[(r/r_0), z]$ represents the intensity of the beam we have

$$A_0^2[(r/r_0), z]2\pi r \, dr = A_0^2[(\rho/r_0), 0]2\pi\rho \, d\rho$$

which on using Eq. (5.87) reduces to

$$A_0^2[(r/r_0), z] = (1/f^2)A_0^2[(r/r_0 f), 0] \tag{5.88}$$

The above equation is identical to Eq. (5.86) when

$$A_0^2[(\rho/r_0), 0] = E_0^2 \mathcal{F}(\rho/r_0) \tag{5.89}$$

represents the radial intensity distribution of the beam at $z = 0$. Thus we may state that if f is determined by using Eq. (5.86) the rays are described by Eq. (5.87).

It is seen from Eq. (5.86) that the intensity on the axis, i.e., at $r = 0$ increases as f decreases; hence the minimum value of f^2 corresponds to the focus. This interpretation is also justified from Eq. (5.87) because the minimum value of f^2 corresponds to the minimum cross section of the beam.

Now let us assume the initial intensity distribution to be Gaussian, i.e.,

$$A_0^2(r, z = 0) = E_0^2 \exp(-r^2/w_0^2) \tag{5.90}$$

The intensity distribution in the medium, using Eq. (5.86), would be given by

$$A_0^2(r, z) = \frac{E_0^2}{f^2(z)} \exp\left[-\frac{r^2}{w_0^2 f^2(z)}\right] \tag{5.91}$$

When the intensity distribution is given by Eq. (5.91), the diffraction term in Eq. (5.75) is given by

$$\frac{1}{k^2 A_0}\left(\frac{\partial^2 A_0}{\partial r^2} + \frac{1}{r}\frac{\partial A_0}{\partial r}\right) = -\frac{2}{k^2 w_0^2 f^2} + \frac{r^2}{k^2 w_0^4 f^4} \tag{5.92}$$

and hence Eq. (5.75) simplifies to

$$2\frac{\partial S}{\partial z}+\frac{2S}{k}\frac{dk}{dz}+\left(\frac{\partial S}{\partial r}\right)^{2} = -\frac{K_2(z)}{K_0(z)}r^2-\frac{2}{k^2w_0^2f^2}+\frac{r^2}{k^2w_0^4f^4} \tag{5.93}$$

It is important to mention here that the Gaussian distribution of intensity is realized when the laser is operating in the TEM_{00} mode and, indeed, most of the lasers do operate in this mode. Hence Eq. (5.93) is applicable to a large number of experimental situations. For non-Gaussian intensity distributions one can obtain a series solution (Sodha et al., 1972).

Substituting for S from Eq. (5.83) in Eq. (5.93) we obtain

$$2\left[\frac{r^2}{2}\frac{d\beta}{dz}+\frac{d\phi(z)}{dz}\right]+2\left[\frac{r^2}{2}\beta(z)+\phi(z)\right]\frac{1}{k}\frac{dk}{dz}$$

$$+[r\beta(z)]^2 = -\frac{K_2(z)}{K_0(z)}r^2-\frac{2}{k^2w_0^2f^2}+\frac{r^2}{k^2w_0^4f^4} \tag{5.94}$$

Equating the coefficient of r^2 on both sides of the above equation, we obtain

$$\frac{1}{f}\frac{d^2f}{dz^2}+\frac{1}{f}\frac{df}{dz}\left(\frac{1}{k}\frac{dk}{dz}\right) = -\frac{K_2(z)}{K_0(z)}+\frac{1}{k^2w_0^4f^4} \tag{5.95}$$

A similar equation for ϕ can be obtained by equating terms independent of r on both sides of Eq. (5.94); however, the solution for ϕ will not be needed by us. For a medium in which K_0 is independent of z, Eq. (5.95) reduces to

$$\frac{1}{f}\frac{d^2f}{dz^2} = -\frac{K_2(z)}{K_0}+\frac{1}{k^2w_0^4f^4} \tag{5.96}$$

It should be noted that, in the geometrical optics approximation ($\lambda \to 0$, therefore $k \to \infty$), the second term on the RHS of Eq. (5.75) disappears and the treatment is independent of the intensity distribution at $z = 0$.

5.2.1. Solution of Eq. (5.96) when K_2 is Independent of z

The first term in the RHS of Eq. (5.96) is due to inhomogeneities in the medium, whereas the second term is due to the effect of diffraction. As such we may consider the following specific cases.

Geometrical Optics Approximation ($\lambda \to 0$)

The term $(1/k^2w_0^4f^4)$ may be neglected and one can readily obtain the solution

$$f=\frac{1}{\cos\alpha}\cos(\zeta-\theta) \tag{5.97}$$

where*

$$\zeta = \left(\frac{K_2}{K_0}\right)^{1/2} z = \delta z$$

and

$$\theta = \tan^{-1}\left[\frac{1}{R}\left(\frac{K_0}{K_2}\right)^{1/2}\right]$$

In writing Eq. (5.97) we have used the boundary conditions

$$\left.\begin{array}{l} f = 1 \\ \dfrac{1}{f}\dfrac{df}{dz} = \dfrac{1}{R} \end{array}\right]_{\text{at } z=0} \tag{5.98}$$

which means that the beam has a width w_0 [if we put $f = 1$ in Eq. (5.91)] and the radius of curvature of wave front is R at $z = 0$. The intensity distribution is therefore of the form

$$I = \frac{I_0}{\sec^2\theta \cos^2(\zeta-\theta)} \exp\left[-\frac{r^2}{w_0^2 \sec^2\theta \cos^2(\zeta-\theta)}\right] \tag{5.99}$$

For $\zeta = [\theta + (2n+1)\pi/2]$, where $n = 0, 1, \ldots$; $f = 0$ and the intensity is infinite on the axis (i.e., at $r = 0$) and zero everywhere else. These points, therefore, correspond to focal points. The distance between two consecutive focal points is thus $\pi(K_0/K_2)^{1/2}$. The system behaves as an oscillatory waveguide. For an incident plane wave $R = \infty$ and $\theta = 0$, therefore, the focal length is simply $(\pi/2)(K_0/K_2)^{1/2}$. For a typical SELFOC fiber fabricated by Uchida *et al.* (1970), $K_2/K_0 \approx 0.2$ mm^{-2} (see Appendix B); thus the focal length is approximately 3.5 mm. Such short focal length lenses can be put to great practical use (Uchida *et al.*, 1970).

Diffraction Dominated Propagation

Next, we consider the other extreme, in which the effect of in-homogeneities is small so that we can neglect the term $-K_2/K_0$ in Eq. (5.96):

$$\frac{1}{f}\frac{d^2f}{dz^2} = \frac{1}{k^2 w_0^4 f^4} \tag{5.100}$$

*We are running out of symbols; ζ defined here is different from the ζ defined in Sec. 5.1 [see Eq. (5.30)].

Multiplying both sides by $2f(df/dz)\,dz$, we obtain

$$2\frac{df}{dz}\frac{d^2f}{dz^2}dz = \frac{2}{k^2 w_0^4 f^3}\frac{df}{dz}dz$$

which on integration gives

$$\left(\frac{df}{dz}\right)^2 = -\frac{1}{k^2 w_0^4 f^2} + \left[\frac{1}{k^2 w_0^4} + \left(\frac{1}{R}\right)^2\right] \tag{5.101}$$

where we have used the condition that when $f = 1$, $df/dz = 1/R$ [see Eq. (5.98)]. Thus

$$\int_1^f \left\{\frac{[1+(k^2 w_0^4/R^2)]f^2 - 1}{k^2 w_0^4 f^2}\right\}^{1/2} df = \pm\int_0^z dz \tag{5.102}$$

On carrying out the integration and subsequent simplification we get

$$f^2 = \frac{1}{[1+(kw_0^2/R)^2]} + \frac{[1+(R/kw_0^2)^2]}{R^2}\left[z + \frac{R}{(R/kw_0^2)^2+1}\right]^2 \tag{5.103}$$

In order to give a physical interpretation of Eq. (5.103), we note that for an incident plane wave front

$$f^2 = 1 + \frac{z^2}{k^2 w_0^4} \tag{5.104}$$

indicating that the beam diameter would continuously increase. This is simply the effect of diffraction. The spreading increases with increase in the wavelength. (Notice that there are no dark rings like the ones found in the Airy pattern.)

Wave Optics Solution

The solution of Eq. (5.96), when both terms on the RHS are retained, can be obtained in a similar manner. The solution is of the form

$$f^2 = \tfrac{1}{2}[(1+C+\xi^2) + \varepsilon\,\sin(2\zeta+\varphi)] \tag{5.105}$$

where

$$C = \frac{K_0}{K_2}\frac{1}{k^2 w_0^4}$$

$$\xi = \left(\frac{K_0}{K_2}\right)^{1/2}\frac{1}{R}$$

$$\varepsilon = [(1-C)^2 + \xi^4 + 2\xi^2(1+C)]^{1/2}$$

$$\sin\varphi = (1/\varepsilon)[1-(C+\xi^2)]$$

$$\cos\varphi = (2/\varepsilon)\xi$$

For an incident plane wave, i.e., $R = \infty$, one readily obtains

$$f^2 = \tfrac{1}{2}[(1+C)+(1-C)\cos 2\zeta] \tag{5.106}$$

If we substitute the above expression for f^2 in Eq. (5.91), we obtain an expression identical to Eq. (5.38) with C replaced by $1/\sigma^4$. Thus, when K_0 and K_2 weakly depend on z, the present approach is expected to give accurate results. Further, when $C = 1$, f^2 is always unity and the beam diameter always remains the same. Physically, this corresponds to the situation when diffraction divergence is exactly balanced by focusing owing to the inhomogeneity in the medium. Notice that $C = 1$ corresponds to

$$w_0 = \left[\frac{c^2}{\omega^2 K_2}\right]^{1/4} \tag{5.107}$$

which corresponds to the fundamental mode.

For $C > 1$ the beam diameter would initially increase until $\zeta = (\pi/4)$[i.e., $z = (\pi/4)(K_0/K_2)^{1/2}$] and then would start decreasing. Thus, the initial beam diameter is so small that the diffraction divergence dominates until the "focusing phenomenon" takes over. For $C < 1$, the beam diameter would decrease right from the beginning. The minimum value of f^2 is C or 1 corresponding to $C < 1$ and $C > 1$, respectively. This is to be compared with the results obtained in the geometrical optics approximation where $(f^2)_{min}$ was always zero, implying infinite intensity on the axis. A finite value of f^2 (at the focal point) is to be expected owing to diffraction effects. The distance between two consecutive focal points is again $\pi(K_0/K_2)^{1/2}$.

We now give the results of some numerical calculations corresponding to the following values of various parameters:

$$w_0 = 100 \, \mu = 10^{-2} \, \text{cm}$$

$$K_2 = 0.1 \, \text{cm}^{-2} \, \text{(constant)}$$

$$\lambda = 5 \times 10^{-5} \, \text{cm}$$

$$K_0 = 2.0$$

The intensity distributions, as obtained from Eq. (5.91) with f^2 given by Eq. (5.106), are plotted in Figs. 5.3a and 5.3b which correspond to $\rho(= r/w_0) = 0$ and 1.483, respectively. For the second value of ρ, the intensity at $z = 0$ is 1/9 the value of intensity at $\rho = 0$.

First we note that the intensity distribution on the axis (i.e., for $\rho = 0$) has very sharp maxima for $z = 7$ cm, 21 cm, etc. These are the positions of the foci. In geometrical optics approximation ($\lambda \to 0$), the intensity would have been infinite at the foci. On the other hand, for other values of ρ the intensity has minima at the foci. This is to be expected because at the foci the intensity is concentrated near the axis.

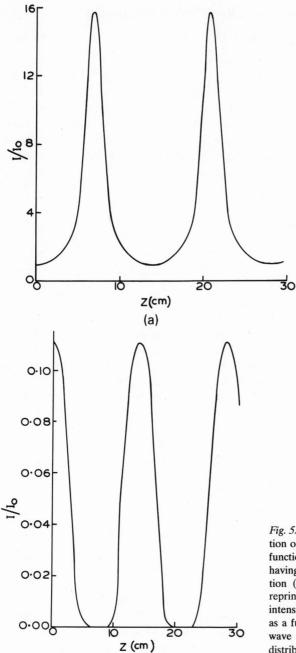

Fig. 5.3. (a) The intensity distribution on the axis (i.e., at $\rho = 0$ as a function of z for an incident wave having Gaussian intensity distribution (after Sodha *et al.*, 1972b; reprinted by permission). (b) The intensity distribution (at $\rho = 1.483$) as a function of z for an incident wave having Gaussian intensity distribution (after Sodha *et al.*, 1972b; reprinted by permission).

Numerical solutions of Eq. (5.96) for different profiles of $K_2(z)$ have been given by Sodha *et al.* (1971b).

5.2.2.   Analytical Solutions for Specific Dependence of $K_2(z)$ and Path of Rays in Conical SELFOC Fibers

We first consider the case when

$$K_2(z) = \frac{K_{20}}{(1+\alpha z)^2} \tag{5.108}$$

For the above dielectric constant variation, Eq. (5.96) becomes

$$\frac{1}{f}\frac{d^2 f}{dz^2} = -\frac{K_{20}}{K_0(1+\alpha z)^2} + \frac{1}{k^2 w_0^4 f^4}$$

or,

$$\frac{1}{f}\frac{d^2 f}{d\xi^2} = -\frac{1}{\beta^2 \xi^2} + \frac{\Lambda}{f^4} \tag{5.109}$$

where

$$\xi = (1+\alpha z)$$

$$\beta^2 = \frac{K_0 \alpha^2}{K_{20}} \tag{5.110}$$

$$\Lambda = \frac{1}{k_0^2 w_0^4 \alpha^2}$$

We introduce the variable

$$F(\xi) = \frac{1}{\xi^{1/2}} f(\xi) \tag{5.111}$$

Thus

$$\frac{df}{d\xi} = \xi^{1/2}\frac{dF}{d\xi} + \tfrac{1}{2}\xi^{-1/2}F(\xi)$$

and

$$\frac{d^2 f}{d\xi^2} = \xi^{1/2}\frac{d^2 F}{d\xi^2} + \xi^{-1/2}\frac{dF}{d\xi} - \tfrac{1}{4}\xi^{-3/2}F(\xi)$$

Substituting in Eq. (5.109) we get

$$\xi\frac{d^2 F}{d\xi^2} + \frac{dF}{d\xi} = -\frac{\gamma^2}{\xi}F + \frac{\Lambda}{\xi F^3} \tag{5.112}$$

where

$$\gamma^2 = (1/\beta^2) - \tfrac{1}{4}$$

and for most practical fibers $\gamma^2 > 0$. In terms of the independent variable

$$\eta = \ln \xi \qquad (5.113)$$

Eq. (5.111) becomes

$$\frac{d^2 F}{d\eta^2} = -\gamma^2 F + \frac{\Lambda}{F^3} \qquad (5.114)$$

If we multiply the above equation by $2(dF/d\eta)\, d\eta$ and integrate, we would obtain

$$\left(\frac{dF}{d\eta}\right)^2 = K - \left(\frac{\Lambda}{F^2} + \gamma^2 F^2\right) \qquad (5.115)$$

where

$$K = \left[(1/\alpha R) - \tfrac{1}{2}\right]^2 + (\Lambda + \gamma^2)$$

and we have used the boundary conditions that

$$f(z)|_{z=0} = 1$$

and

$$\frac{1}{f}\frac{df}{dz}\bigg|_{z=0} = \frac{1}{R} \qquad (5.116)$$

Since $z = 0$ corresponds to $\xi = 1$ (i.e., $\eta = 0$), Eq. (5.114) implies

$$F(\eta)|_{\eta=0} = 1$$

and

$$\frac{dF}{d\eta}\bigg|_{n=0} = \frac{1}{\alpha R} - \frac{1}{2} \qquad (5.117)$$

Equation (5.115) can be integrated by introducing a new variable

$$\mathscr{F}(\eta) \equiv \gamma F^2 - (K/2\gamma) \qquad (5.118)$$

the result of the integration is

$$\mathscr{F} = \left(\frac{1}{\alpha R} - \frac{1}{2}\right)\sin(2\gamma\eta) + \left(\gamma - \frac{K}{2\gamma}\right)\cos(2\gamma\eta)$$

Using Eqs. (5.111), (5.118), and (5.113) we finally obtain

$$f^2(\zeta) = (1 + \beta\zeta)\left\{\left[\frac{K}{2\gamma^2} + \frac{1}{\gamma}\left(\frac{1}{\alpha R} - 2\right)\right]\sin[2\gamma\,\ln(1 + \beta\zeta)]\right.$$

$$\left. + \left(1 - \frac{K}{2\gamma^2}\right)\cos[2\gamma\,\ln(1 + \beta\zeta)]\right\} \qquad (5.119)$$

where

$$\beta = (K_0/K_{20})^{1/2}\alpha$$

and

$$\zeta = (K_{20}/K_0)^{1/2}z$$

For an incident plane wave ($R = \infty$) one obtains

$$f^2(\zeta) = (1 + \beta\zeta)\left\{\left[\frac{1}{2} + \frac{1}{\gamma^2}\left(\frac{1}{8} + \frac{\Lambda}{2}\right)\right] - \frac{1}{2\gamma}\sin[2\gamma\,\ln(1 + \beta\zeta)]\right.$$

$$\left. + \left[\frac{1}{2} - \frac{1}{\gamma^2}\left(\frac{1}{8} + \frac{\Lambda}{2}\right)\right]\cos[2\gamma\,\ln(1 + \beta\zeta)]\right\} \qquad (5.120)$$

Equations (5.119) and (5.120) give the variation of the dimensionless beamwidth parameter and will be used in Chapter 8 to study mode conversion.

In the geometrical optics approximation $\Lambda \to 0$ and one obtains after some simple manipulations

$$f^2(\zeta) = (1 + \beta\zeta)\{\cos[\gamma\,\ln(1 + \beta\zeta)] - \frac{1}{2\gamma}\sin[\gamma\,\ln(1 + \beta\zeta)]\}^2 \qquad (5.120a)$$

The above solutions correspond to $\beta < 2$ (i.e., $\gamma^2 > 0$). For $\beta > 2$ one obtains hyperbolic functions; for example, instead of Eq. (5.120a) one gets

$$f^2(\zeta) = (1 + \beta\zeta)\{\cosh[\gamma\,\ln(1 + \beta\zeta)] - \frac{1}{2\gamma}\sinh[\gamma\,\ln(1 + \beta\zeta)]\}^2 \qquad (5.120b)$$

where

$$\gamma = \left|\frac{1}{4} - \frac{1}{\beta^2}\right|^{1/2}$$

We next consider the case when

$$K_2(z) = K_{20}(1 + \alpha z) \qquad (5.121)$$

For the above dielectric constant variation, an analytical solution of Eq. (5.96) can be obtained only in the geometrical optics approximation (i.e., $k \to \infty$). Under such approximation, Eq. (5.96) can be transformed to the following form

$$\frac{d^2f}{d\xi^2} + \xi f(\xi) = 0 \tag{5.122}$$

where

$$\xi = \beta^{1/2}\zeta + \frac{1}{\beta} \qquad \text{for} \qquad \alpha > 0$$

$$= -\beta^{1/2}\zeta + \frac{1}{\beta} \qquad \text{for} \qquad \alpha < 0$$

$$\beta = [(K_0/K_{20})^{1/2}|\alpha|]^{2/3}$$

and

$$\zeta = (K_{20}/K_0)^{1/2}z$$

The solution of Eq. (5.122), subject to the boundary condition represented by Eq. (5.116) (with $R = \infty$), is given by

$$f(\xi) = \frac{U_2'(\beta^{-1})U_1(\xi) - U_1'(\beta^{-1})U_2(\xi)}{U_2'(\beta^{-1})U_1(\beta^{-1}) - U_2(\beta^{-1})U_1'(\beta^{-1})} \tag{5.123}$$

where

$$U_1(\xi) = \frac{\Gamma(\frac{2}{3})}{3^{1/3}}\xi^{1/2}J_{-1/3}(\tfrac{2}{3}\xi^{3/2})$$

and

$$U_2(\xi) = 3^{1/3}\Gamma(\tfrac{4}{3})\xi^{1/2}J_{1/3}(\tfrac{2}{3}\xi^{3/2})$$

The functions $U_1(\xi)$ and $U_2(\xi)$ along with their derivatives have been tabulated by Smirov (1960).

An analytical solution can also be obtained when

$$K_2(z) = K_{20}(1 + \alpha_1 e^{\beta z} + \alpha_2 e^{2\beta z})$$

The solution is in terms of hypergeometric functions (see, for example, Murphy, 1960). Further, with periodic variation of $K_2(z)$ such that

$$K_2(z) = K_{20}(1 + \alpha \cos 2\beta z)$$

the resulting equation becomes Mathieu's equation, the solution of which is given in many places (see, for example, Margenau and Murphy, 1956).

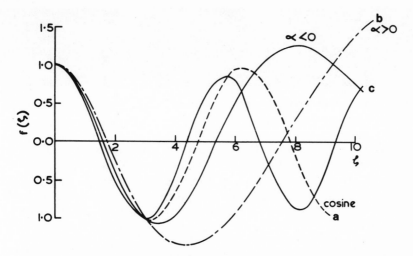

Fig. 5.4. Variation of $f(\zeta)$ with $\zeta [=(K_{20}/K_0)^{1/2}z]$ for various forms of $K_2(z)$ in the geometrical optics approximation. Curves (a), (b), and (c) correspond to $K_2(z)=K_{20}$, Eq. (5.108), and Eq. (5.121), respectively. $\beta=0.2$ (after Sodha *et al.*, 1971a; reprinted by permission).

In Fig. 5.4 we present the variation of $f(\zeta)$ with ζ for various values of α. Whenever $|f(\zeta)|$ increases with z the beam will diverge and vice versa.

In a recent paper, Kita and Uchida (1969) have reported the fabrication of a conical SELFOC fiber that has a radially parabolic variation of refractive index at any cross section perpendicular to the fiber axis, and with $a_2(z)$ varying gradually along the fiber axis; $a_2(z)$ is defined through the equation

$$n = n_0(z)[1 - a_2(z)r^2] \qquad (5.124)$$

where $n(r, z)$ represents the refractive index variation. Thus

$$K_0(z) = n_0^2(z)$$

and

$$K_2(z) = 2n_0^2(z)a_2(z) \qquad (5.125)$$

An incident ray may pass through the SELFOC fiber without hitting the side wall of the guide (see Fig. 5.5). The refractive index at the periphery of the guide does not change with the external shape of the fiber. Thus, $a_2(z)$ can be estimated from the external shape of the fiber by

$$a_2(z) = \frac{\Delta n}{n_0 R^2(z)} \qquad (5.126a)$$

Fig. 5.5. A laser beam passing through a conical SELFOC fiber (after Kita and Uchida, 1969; reprinted by permission).

where $R(z)$ is the variation of the radius of the fiber and Δn is the difference between the index on the axis and the periphery. The variation of $a_2(z)$, using the above equation, is plotted (as the solid curve) in Fig. 5.6. The values of Δn and n_0 are 0.025 and 1.6 respectively. Also plotted in the figure is the variation of $a_2(\zeta)$ as obtained from the following equation [cf. Eq. (5.118)]:

$$a_2(z) = \frac{a_{20}}{(1 - 0.03\zeta)^2} \qquad (5.126b)$$

where

$$\zeta = (2a_{20})^{1/2} z = (K_{20}/K_0)^{1/2} z$$

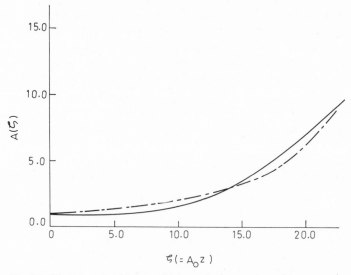

Fig. 5.6. Variation of $a_2(\zeta)/a_{20}$ with ζ; a_{20} is the value of $a_2(\zeta)$ at $\zeta = 0$. Solid and dashed curves correspond to Eqs. (5.124) and (5.125), respectively (after Ghatak et al., 1972; reprinted by permission).

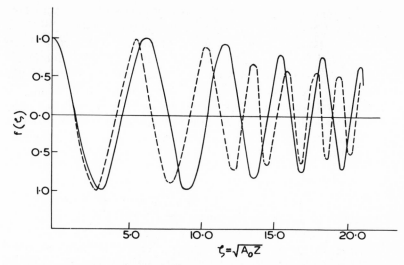

Fig. 5.7. Variation of the dimensionless beamwidth parameter, $f(\zeta)$ with ζ. The solid curve corresponds to the calculation using Eq. (5.119). The dashed curve is the experimental curve (after Ghatak *et al.*, 1972; reprinted by permission).

It can be seen that the guide fabricated by Kita and Uchida (1969) is very nearly conical. Using the variation of $a_2(z)$ as given by Eq. (5.126a), the calculated variation of $f(z)$ is plotted in Fig. 5.7 and compared with the experimental data. The agreement is fairly satisfactory.

We may conclude that the formulation developed in Sec. 5.2 can be used successfully in describing the propagation of laser beams through SELFOC fibers.

5.2.3.   Resolving Power of a SELFOC Lens

The resolving power of a SELFOC lens was investigated under the configuration shown in Fig. 5.8 (Uchida *et al.*, 1970). The experimental results for the resolving power under white light illumination comes out to be about 450 lines/mm for a SELFOC lens with length 3.34 mm, diameter 1.0 mm, and focal length, $F = 1.9$ mm. Thus $n_0 = 1.63$ and $a_2 = 0.0615$ mm^{-2} [a_2 and n_0 have been defined in Eq. (5.124)]. The distance from the exit aperture of the SELFOC lens to the focal point, d, is given by (Miller, 1965; Ghatak *et al.*, 1972):

$$d = \frac{1}{n_0(2a_2)^{1/2}} \cot[(2a_2)^{1/2}t]$$

$$= 0.74 \text{ mm}$$

where t is the length of the SELFOC lens (here $t = 3.34$ mm). The above value of d agrees very well with the experimental configuration (Fig. 5.8).

Further, one can show that the diameter of the beam at the exit aperture will be $\alpha \cos(2a_2)^{1/2}t$, where α is the diameter of the beam at the entrance of the guide. The experimental configuration corresponds to $\alpha = 1.0$ mm, thus the diameter of the beam at the exit aperture turns out to be 0.3895 mm. The image pattern will now be identical to the Airy pattern formed by a plane wave incident on a lens of diameter 0.3895 mm with focal length 0.74 mm. The cutoff spatial frequency in lines/mm of such a system will be given by (Goodman, 1968)

$$\Omega = \frac{\omega}{2\lambda F^*}$$

where F^* (known as the f number) is the ratio of the focal length of the lens to its diameter, ω is the dimensionless spatial frequency, and λ is the wavelength of radiation in mm. It is well known that the cutoff value of ω is 2.0. Therefore, for the present case, using $\lambda = 6.4 \times 10^{-4}$ mm, the cutoff frequency comes out to be $\Omega = 880$ lines/mm, which is roughly twice the value reported by Uchida *et al.* (1970). This discrepancy has been discussed by Ghatak *et al.* (1972) and may be attributed to the following:

(i) The light used in the experimental arrangement was white light which may tend to decrease the resolving power.

(ii) Further, in the calculations the maximum value of ω (=2) has been used, which corresponds to zero contrast. Any experimental arrangement would correspond to a value of ω less than 2.

Assuming a blackbody source at 3000°K and photopic vision the number of lines/mm for a contrast threshold of 7%–8% comes out to be approximately 750. The remaining difference between the experimental and theoretical values can be partly attributed to the irregularity of parabolic

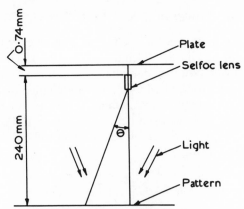

Fig. 5.8. Configuration for resolving power measurement (after Uchida *et al.*, 1970).

refractive index variation, insufficient flatness of the terminal surfaces, and the approximations made in the theory. Further, the fact that the test target used in the experiment was a square-wave three-bar target instead of a sine wave, will make a difference of not more than 50 lines/mm at the specified contrast threshold.

5.2.4. Focusing by a Gas Lens

Gas lenses are also characterized by parabolic refractive index variation (similar to SELFOC fibers), and have been suggested as focusing elements of a beam waveguide. In Appendix B we calculate the temperature variation and the corresponding refractive index variation inside a gas lens. It should be pointed out right in the beginning that, whereas the z dependence of n_0 is not very important, the variation of a_2 with z is very rapid and should be taken into account for quantitative comparison with the experimental data. Hence, the modal analysis approach is quite unsuitable for the present study.

In this section we outline the approach of Ghatak *et al.* (1973). In general, the effects due to diffraction are very small, and Eq. (5.95) in the geometrical optics approximation simplifies to

$$\frac{1}{f}\frac{d^2f}{dz^2}+\left(\frac{1}{n_0}\frac{dn_0}{dz}\right)\frac{1}{f}\frac{df}{dz}=-2a_2(z) \qquad (5.127)$$

Assuming n_0 and $a_2(z)$ to be given by Eqs. (B.17) and (B.18) a rigorous solution of Eq. (5.127) can be obtained, which is given by

$$f=\frac{C_1-C_2\ln\xi}{C_3}\sum_{r=0}^{\infty}\xi^r(b_r)_{s=0}-\frac{C_2}{C_3}\sum_{r=0}^{\infty}\xi^r\left(\frac{db_r}{ds}\right)_{s=0} \qquad (5.128)$$

where

$$\xi=\exp(-4\alpha z)$$

$$C_1=\sum_{r=0}^{\infty}(b_r)_{s=0}+\sum_{r=0}^{\infty}r\left(\frac{db_r}{ds}\right)_{s=0}$$

$$C_2=\sum_{r=0}^{\infty}r(b_r)_{s=0}$$

$$C_3=C_1\sum_{r=0}^{\infty}(b_r)_{s=0}-C_2\sum_{r=0}^{\infty}\left(\frac{db_r}{ds}\right)_{s=0} \qquad (5.129)$$

$$\frac{b_{r+1}}{b_r}=-\frac{B_1}{n_{00}-B_1}\frac{(r+s)(r+s-1)+(8\alpha^2a^2)^{-1}}{(r+s+1)^2} \qquad \text{for } r=0,1,2,\ldots$$

$$b_0=1$$

$$B_1=(n_{00}-1)\frac{T_s-T_{00}}{T_{00}}$$

and other symbols have been defined in Appendix B. For values of various parameters in a typical experimental arrangement, the infinite series is very rapidly convergent and a few terms would suffice to give fairly accurate results.

The focal length of the gas lens is obtained by calculating the radius of curvature of the wave front $\{ = [(1/f)(df/dz)]^{-1}\}$ at the exit aperture of the lens.

In Figs. 5.9 and 5.10 the theoretical results have been compared with the experimental data of Kaiser (1968a) and Aoki and Suzuki (1967) for the focal length in a tubular gas lens. The following values of various parameters have been used:

$$L = 15 \text{ cm}$$

$$a = 0.317 \text{ cm} \qquad \text{for the experimental arrangement of Kaiser (1968)}$$

$$\frac{\kappa}{\rho C_p} = 0.177 \text{ cm}^2/\text{sec}$$

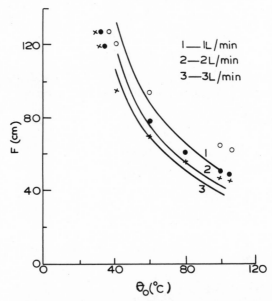

Fig. 5.9. $\bigcirc$, $\bullet$, and $\times$ represent the experimental data (corresponding to flowrates of 1, 2, and 3 liters/min, respectively) of Kaiser (1968a) for the focal length as a function of $\theta_0 (= T_s - T_{00})$ for a gas lens using air. The solid curves correspond to the theoretical results using Eqs. (B.17) and (B.18) for the refractive index variation (after Ghatak *et al.*, 1973; reprinted by permission).

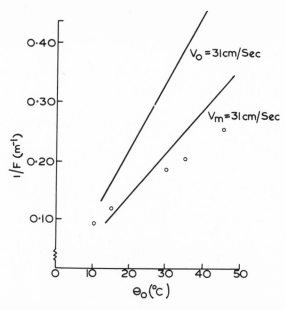

Fig. 5.10. The circles represent the experimental data of Aoki and Suzuki (1967) for the reciprocal focal length of a gas lens (using air) as a function of θ_0 ($= T_s - T_{00}$). The curves correspond to theoretical results using Eqs. (B.17) and (B.18) for the refractive index variation (after Ghatak *et al.*, 1973; reprinted by permission).

and

$$L = 45 \text{ cm}$$

$$a = 0.85 \text{ cm} \qquad \text{for the experimental arrangement of Aoki and Suzuki (1967)}$$

$$\frac{\kappa}{\rho C_p} = 0.177 \text{ cm}^2/\text{sec}$$

In the experimental arrangement of Aoki and Suzuki (1967) the reported wind velocity was 0.31 m/sec. But it was not mentioned whether it corresponds to v_0 or v_m, where $v_0 = 2v_m$ is the velocity of the gas at the axis of the lens. Results corresponding to two sets of calculations, one corresponding to $v_0 = 0.31$ m/sec and other corresponding to $v_m = 0.31$ m/sec are shown in Fig. 5.10. We find that the agreement with the experimental results is fairly satisfactory.

Corresponding to the experimental data given by Eq. (5.130b) we have tried to verify the thin-lens formula

$$\frac{1}{v} + \frac{1}{u} = \frac{1}{F},$$

where u is the distance of the object from the entrance aperture, v is the distance of the image from the exit aperture, and F is the focal length. Thus,

$$\left(\frac{1}{f}\frac{df}{dz}\right)_{z=0} = \frac{1}{u}$$

$$\left(\frac{1}{f}\frac{df}{dz}\right)_{z=L} = \frac{1}{v}$$

In Fig. 5.11 we have plotted the calculated values of $uv/(u+v)$ for various values of u. The horizontal line is the value of focal length as obtained by assuming $u = \infty$. We find that the thin-lens formula is satisfied within about 3%.

It should be mentioned that as the flow velocity, v_m, is increased from very low values initially, the temperature gradient will increase until the flow tends to be turbulent. With the onset of turbulence further increase of v_m leads to a lower temperature gradient. Hence, the focal length of the gas lens passes through a minimum as the flow velocity is increased; the minimum focal length corresponds to the maximum temperature gradient and is associated with the onset of turbulence.

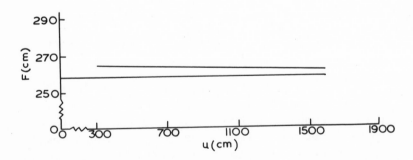

Fig. 5.11. The dependence of the focal length on u corresponding to the refractive index variation as given by Eqs. (B.17) and (B.18). The values of various parameters are the same as given in Eq. (5.130b) with $\theta_0 = 19.5°C$ and $v_m = 31$ cm/sec (after Ghatak *et al.*, 1973; reprinted by permission).

5.2.5. Effect of Random Inhomogeneities in a SELFOC Fiber

Owing to turbulence and unsteady flow in the case of the gas lens, and the limitations of the manufacturing techniques in the case of SELFOC fibers, K_2 is expected to fluctuate in a random manner around a mean value of K_{20}.

In this section, we have assumed that within a small length $\Delta(K_0/K_{20})^{1/2}$, K_2 remains constant and that in two different intervals of length $\Delta(K_0/K_{20})^{1/2}$, there is no correlation in the values of K_2; further, K_2 is assumed to have a normal distribution over the different intervals.

Using this model, we have studied the dependence of focusing properties on Δ (which is a measure of the correlation length) and on the magnitude of standard deviation in K_2.

The random variation of K_2 may be represented by

$$K_2(z) = K_{20}(1 + A(z)) = K_{20}b(z) \qquad (5.131)$$

where $b(z)$ remains constant over intervals of length $\Delta(K_0/K_{20})^{1/2}$ and has a normal distribution with the mean value zero and standard deviation σ.

In the remaining part of this section we have developed a perturbation theory based on the geometrical optics approximation; Goyal *et al.* (1973) have justified this approach by comparison of results obtained on the computer by employing the wave optics approach for a specific typical case with those obtained by the present theory. In this approximation the dimensionless beamwidth parameter satisfies the following equation [Eq. (5.108)].

$$\frac{d^2f}{d\zeta^2} + b(\zeta)f(\zeta) = 0 \qquad (5.132)$$

where

$$\zeta = (K_{20}/K_0)^{1/2}z \qquad (5.133)$$

Following Goyal *et al.* (1973) we use the perturbation theory approach and write

$$K_2(z) = K_{20} + \Lambda K_{2r}(z) \qquad (5.134)$$

and

$$f = f_0 + \Lambda f_1 + \Lambda^2 f_2 + \cdots \qquad (5.135)$$

where $K_{2r} = K_{20}A(z)$ and Λ is a parameter which is set equal to 1 in the final expression. Substituting the above expression in Eq. (5.132) and equating various powers of Λ, we obtain

$$\frac{d^2 f_0}{d\zeta^2} + f_0 = 0$$

$$\frac{d^2 f_1}{d\zeta^2} + f_1 = -A(\zeta)f_0 \qquad (5.136)$$

$$\frac{d^2 f_2}{d\zeta^2} + f_2 = -A(\zeta)f_1$$

Assuming an incident plane wave front, i.e., $df_0/d\zeta = 0$ at $\zeta = 0$, we obtain

$$f_0 = \cos\zeta,$$

$$f_1 = -\tfrac{1}{2}\sin\zeta \int_0^\zeta (1 + \cos 2\zeta')A(\zeta')\,d\zeta' + \tfrac{1}{2}\cos\zeta \int_0^\zeta \sin 2\zeta' A(\zeta')\,d\zeta'$$

$$(5.137)$$

and so on.

In the first-order perturbation theory

$$f - f_0 = \int_0^\zeta \tfrac{1}{2}A(\zeta')[\cos\zeta \sin 2\zeta' - \sin\zeta(1 + \cos 2\zeta')]\,d\zeta' \qquad (5.138)$$

Since $A(\zeta)$ is a constant in the interval $(r-1)\Delta < \zeta < r\Delta$, we may write the above integral as

$$f(\zeta) - f_0(\zeta) = \sum_{r=1}^n A_r F_r(\zeta') \qquad (5.139)$$

where we have divided the region 0–ζ in n intervals, i.e.,

$$\zeta = n\Delta$$

Further, A_r is the value of $A(\zeta')$ in the rth interval, and

$$F_r(\zeta) = \tfrac{1}{2}\cos\zeta \int_{(r-1)\Delta}^{r\Delta} \sin 2\zeta'\,d\zeta' - \tfrac{1}{2}\sin\zeta \int_{(r-1)\Delta}^{r\Delta} (1 + \cos 2\zeta')\,d\zeta' \qquad (5.140)$$

$$\simeq \tfrac{1}{2}\Delta(\cos\zeta \sin[2(r-\tfrac{1}{2})\Delta] - \sin\zeta\{1 + \cos[2(r-\tfrac{1}{2})\Delta]\}) \qquad (5.141)$$

We choose a set of n random numbers* such that

$$\sum_{r=1}^{n} A_r = 0 \tag{5.142}$$

$$\sum_{r=1}^{n} A_r^2 = \sigma^2 n \tag{5.143}$$

In order to obtain different sets, we permute the same numbers at different places, thus obtaining $n!$ sets. We take the average of these $n!$ sets which gives us

$$\langle f - f_0 \rangle = \frac{(n-1)! \sum_{r=1}^{n} A_r \sum_{r=1}^{n} F_r}{n!} = 0 \tag{5.144}$$

at all values of ζ. Further, it may be shown that

$$\langle (f - f_0)^2 \rangle = \left\langle \left(\sum_r A_r F_r \right)^2 \right\rangle$$

$$= \frac{\sigma^2}{(\zeta - \Delta)} \Delta [\zeta I_2(\zeta) - I_1^2(\zeta)] \tag{5.145}$$

where

$$I_1(\zeta) = \tfrac{1}{4} \cos \zeta (1 - \cos 2\zeta) - \tfrac{1}{2} \sin \zeta (\zeta + \sin \zeta \cos \zeta) \tag{5.146}$$

and

$$I_2(\zeta) = \sum_{r=1}^{n} (\tfrac{1}{2} \cos \zeta \sin[(2r-1)\Delta] - \tfrac{1}{2} \sin \zeta \{1 + \cos[(2r-1)\Delta]\})^2 \Delta \tag{5.147}$$

Since, in general, $\zeta \gg \Delta$, this equation shows that $\langle (f - f_0)^2 \rangle$ is directly proportional to the correlation length and to the square of the standard deviation in K_2.

5.3. Considerations for General Inhomogeneous Media

In a recent paper Casperson (1973) has considered propagation of Gaussian light beams through general inhomogeneous media. The starting

*Obviously, ζ should be much greater than Δ, so that n is large.

point is again the scalar wave equation

$$\nabla^2 \psi + k^2 \psi = 0 \qquad (5.148)$$

where k^2 is now assumed to be of the form

$$k^2(x, y, z) = k_0(z)[k_0(z) - k_{1x}(z)x - k_{1y}(z)y$$
$$- k_{2x}(z)x^2 - 2k_{xy}(z)xy - k_{2y}(z)y^2] \qquad (5.149)$$

which represents the most general second-degree profile possible. If we assume the solution of the form

$$\psi = G(x, y, z) \exp\left[-i \int k_0(z)\, dz \right] \qquad (5.150)$$

then $G(x, y, z)$ would satisfy the equation

$$\frac{\partial^2 G}{\partial x^2} + \frac{\partial^2 G}{\partial y^2} - 2ik_0 \frac{\partial G}{\partial z} - i\frac{dk_0}{dz}G - k_0[k_{1x}x + k_{1y}y$$
$$+ k_{2x}x^2 + 2k_{xy}xy + k_{2y}y^2]G(x, y, z) = 0 \qquad (5.151)$$

where G is assumed to vary slowly with z, so that $\partial^2 G/\partial z^2$ can be neglected [cf. Eq. (5.72)]. For an incident Gaussian beam, Casperson (1973) assumes G to be of the form

$$G(x, y, z) = \exp\{-i\,[\tfrac{1}{2}Q_x(z)x^2 + Q_{xy}(z)xy$$
$$+ \tfrac{1}{2}Q_y(z)y^2 + S_x(z)x + S_y(z)y + P(z)]\} \qquad (5.152)$$

The size of the beam and the curvature of the phase fronts are governed by the complex beam parameters Q_x, Q_{xy}, and Q_y. If the above expression for $G(x, y, z)$ is substituted in Eq. (5.151), one finds (by equating equal powers of x, y, and xy) that the beam parameters are governed by the following equations:

$$Q_x^2 + Q_{xy}^2 + k_0\left(\frac{dQ_x}{dz}\right) + k_0 k_{2x} = 0 \qquad (5.153)$$

$$Q_y^2 + Q_{xy}^2 + k_0\left(\frac{dQ_y}{dz}\right) + k_0 k_{2y} = 0 \qquad (5.154)$$

$$(Q_x + Q_y)Q_{xy} + k_0\left(\frac{dQ_{xy}}{dz}\right) + k_0 k_{xy} = 0 \qquad (5.155)$$

$$Q_x S_x + Q_{xy} S_y + k_0 \frac{dS_x}{dz} + \tfrac{1}{2} k_0 k_{1x} = 0 \tag{5.156}$$

$$Q_y S_y + Q_{xy} S_x + k_0 \frac{dS_y}{dz} + \tfrac{1}{2} k_0 k_{1y} = 0 \tag{5.157}$$

$$\frac{dP}{dz} = -i \left[\frac{1}{2k_0} (Q_x + Q_y) \right] - \frac{1}{2k_0} (S_x^2 + S_y^2) - \frac{i}{2k_0} \frac{dk_0}{dz} \tag{5.158}$$

Thus for an incident Gaussian beam, one first makes a Taylor series expansion of $k(x, y, z)$ (about the center of the beam) to obtain the coefficients $k_0(z)$, $k_{1x}(z)$, $k_{1y}(z)$, $k_{xy}(z)$, etc. Once these coefficients are known one can solve the six equations (5.153)–(5.158) to obtain the six functions Q_x, Q_{xy}, Q_y, S_x, S_y, and P, and thereby obtain the complete solution for the problem. This procedure may, in general, be difficult but numerical solutions are always possible. However, if the symmetry axes of the quadratic approximation to the medium remain parallel to the x and y axes then $k_{xy} = 0$; and, if the symmetry axes of the astigmatic beam are also initially parallel to the x and y axes (i.e., $Q_{xy}(0) = 0$) then [from Eq. (5.155)] one readily obtains $Q_{xy}(z) = 0$. Thus Eqs. (5.153)–(5.158) simplify to

$$Q_x^2 + k_0 \frac{dQ_x}{dz} + k_0 k_{2x} = 0 \tag{5.159}$$

$$Q_y^2 + k_0 \frac{dQ_y}{dz} + k_0 k_{2y} = 0 \tag{5.160}$$

$$Q_x S_x + k_0 \frac{dS_x}{dz} + \tfrac{1}{2} k_0 k_{1x} = 0 \tag{5.161}$$

$$Q_y S_y + k_0 \frac{dS_y}{dz} + \tfrac{1}{2} k_0 k_{1y} = 0 \tag{5.162}$$

$$\frac{dP}{dz} = -i \frac{1}{2k_0} (Q_x + Q_y) - \frac{1}{2k_0} (S_x^2 + S_y^2) - \frac{i}{2k_0} \frac{dk_0}{dz} \tag{5.163}$$

and the Gaussian beam has the form

$$G(x, y, z) = \exp[-i(\tfrac{1}{2} Q_x x^2 + \tfrac{1}{2} Q_y y^2 + S_x x + S_y y + P)] \tag{5.164}$$

If the beam parameters are separated into their real and imaginary parts

$$Q_x = Q_{xr} + iQ_{xi}$$
$$S_x = S_{xr} + iS_{xi}$$
$$P = P_r + iP_i$$

then Eq. (5.164) takes the form

$$G = \exp\left\{-i\left[\frac{Q_{xr}}{2}(x - d_{xp})^2 + \frac{Q_{yr}}{2}(y - d_{yp})^2 - \tfrac{1}{2}Q_{xr}d_{xp}^2 - \tfrac{1}{2}Q_{yr}d_{yp}^2 + P_r\right]\right.$$
$$\left. + \left[\tfrac{1}{2}Q_{xi}(x - d_{xa})^2 + \tfrac{1}{2}Q_{yi}(y - d_{ya})^2 - \tfrac{1}{2}Q_{xi}d_{xa}^2 - \tfrac{1}{2}Q_{yi}d_{ya}^2 + P_i\right]\right\} \qquad (5.165)$$

where $d_{xa} = -S_{xi}/Q_{xi}$ and $d_{xp} = -S_{xr}/Q_{xr}$ are the displacements in the x direction of the center of the amplitude distribution and of the phase front, respectively.

If we assume k_0, k_{1x}, k_{1y}, k_{2x}, and k_{2y} to be constants, analytical solutions can be readily found. The solution of Eq. (5.159) can be written in the form

$$\frac{1}{k_0}Q_x(z) = \frac{1}{F_x}\frac{dF_x}{dz} \qquad (5.166)$$

where

$$F_x(z) = \cos\left[\left(\frac{k_{2x}}{k_0}\right)^{1/2}z\right] + \frac{Q_x(0)}{k_0}\left(\frac{k_0}{k_{2x}}\right)^{1/2}\sin\left[\left(\frac{k_{2x}}{k_0}\right)^{1/2}z\right] \qquad (5.167)$$

Further, using Eq. (5.162) it can be shown that $S_x'[= S_x - k_{1x}Q_x/(2k_{2x})]$ satisfies the equation

$$\frac{dS_x'}{dz} + \frac{Q_x}{k_0}S_x' = 0 \qquad (5.168)$$

the solution of which is

$$S_x'(z)\left[= S_x - \frac{k_{1x}}{2k_{2x}}Q_x\right] = \frac{S_x'(0)}{F_x(z)} \qquad (5.169)$$

Similar solutions can be obtained for Q_y and S_y. If we define

$$P' = P - \frac{k_{1x}}{2k_{2x}}S_x - \frac{k_{1y}}{2k_{2y}}S_y + \frac{k_{1x}^2}{8k_{2x}^2}Q_x + \frac{k_{1y}^2}{8k_{2y}^2}Q_y - \frac{k_{1x}^2}{8k_{2x}}z - \frac{k_{1y}^2}{8k_{2y}}z \qquad (5.170)$$

then, using Eq. (5.163), one obtains

$$\frac{dP'}{dz} = -i\left(\frac{Q_x + Q_y}{2k_0}\right) - \frac{1}{2k_0}(S_x'^2 + S_y'^2) \tag{5.171}$$

On solving Eq. (5.166) and using Eqs. (5.169) and (5.171) we get

$$P'(z) - P'(0) = -\frac{i}{2}\ln[F_x(z)F_y(z)] - \frac{[S_x'(0)]^2}{2k_0}\frac{(k_0/k_{2x})^{1/2}\sin(k_{2x}/k_0)^{1/2}z}{F_x(z)}$$

$$-\frac{[S_y'(0)]^2}{2k_0}\frac{(k_0/k_{2y})^{1/2}\sin(k_{2y}/k_0)^{1/2}z}{F_y(z)} \tag{5.172}$$

Thus, the complete solution has been found and, on substituting for Q_x, Q_y, S_x, S_y, and $P(z)$ in Eq. (5.164), one can study the propagation of the incident beam.

We separate the propagation constant into its real and imaginary parts

$$k = \beta + i\alpha \tag{5.173}$$

In Fig. 5.12 the beam displacement $d_{xa}'[=\sqrt{2}(\beta_0\beta_{2x})^{-1/4}d_{xa}]$ is plotted as a function of $z'[=(\beta_{2x}/\beta_0)^{1/2}z]$ for various values of α_{2x}/β_{2x} with $k_{1x} = 0$. It is evident from the figure that even a very weak negative gain profile $(|\alpha_{2x}| \ll \beta_{2x})$ may make the waveguide extemely unstable. More details can be found in the paper by Casperson (1973).

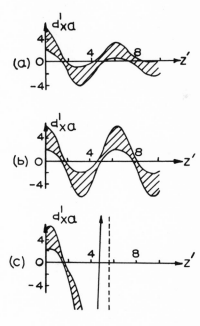

Fig. 5.12. Propagation of an off-axis beam showing the normalized beam displacement in the x direction as a function of z' for (a) $\alpha_{2x}/\beta_{2x} = 0.1$, (b) $\alpha_{2x}/\beta_{2x} = 0$, and (c) $\alpha_{2x}/\beta_{2x} = -0.1$ (after Casperson, 1973; reprinted by permission).

5.4. Summary

In this chapter we have studied, in some detail, the solution of the scalar wave equation for the propagation of light beams through a square-law medium. The modal analysis approach can be used for arbitrary spatial distribution of the incident beam intensity. The parabolic equation approach, although applicable to restricted forms of incident intensity distribution, gives closed form solutions for arbitrary z dependence of the refractive index. Section 5.3 discusses direct solution for Gaussian beam propagation in a general inhomogeneous medium. The theories developed have been used to explain available experimental data.

CHAPTER 6

Vector Theory for Inhomogeneous Circular Waveguides

In Chapter 4 we obtained rigorous solutions of Maxwell's equation for a cladded fiber characterized by the following dielectric constant variation:

$$K = K_1 \qquad \text{for } 0 < r < a$$
$$= K_2 \qquad \text{for } r > a \qquad\qquad (6.1)$$

We derived the equation which would determine the guided modes of the system. In Chapter 5 we obtained solutions of the scalar wave equation for a fiber which was characterized by the following dielectric constant variation:

$$K = K_0 - K_2 r^2 \qquad\qquad (6.2)$$

In this chapter we will solve Maxwell's equations for a radially inhomogeneous system.* We will also discuss various approximate methods for obtaining solutions where rigorous solutions are not possible. These will include the WKB method, perturbation methods, and the variational method.

*We follow the treatment of Kurtz and Streifer (1969a); the cylindrical system of coordinates is used.

6.1. General Solutions for Arbitrary Inhomogeneity

The permittivity variation will be assumed to be of the form

$$\varepsilon = \varepsilon_1[1 - h(r)] \tag{6.3}$$

with $h(0) = 0$. Thus ε_1 is the relative permittivity at the guide center. The function $h(r)$ is an analytic, monotonically increasing function satisfying the relation

$$0 < h(r) \ll 1 \qquad \text{for } 0 < r < a \tag{6.4}$$

where a represents the radius of the core. Thus the permittivity is assumed to decrease monotonically as a function of the distance from the axis with a small fractional change in its value as we go from the axis to the boundary of the core. The corresponding dielectric constant variation will be of the form

$$K = K_0[1 - h(r)] \tag{6.5}$$

where $K_0 = \varepsilon_1/\varepsilon_0$.

In order to carry out a rigorous modal analysis we assume the field components to be of the form (see Sec. 4.4)

$$\mathcal{E}_i(r, \varphi, z; t) = E_i(r) \exp[i(\omega t - \beta z - l\varphi)] \tag{6.6}$$

and

$$\mathcal{H}_i(r, \varphi, z; t) = H_i(r) \exp[i(\omega t - \beta z - l\varphi)] \tag{6.7}$$

where the subscript i stands for any component (i.e., r, φ, or z) of the fields; and for the fields to be single valued l must be $0, \pm 1, \pm 2, \ldots$. The transverse field components (E_φ, E_r, H_φ, and H_r) can easily* be expressed in terms of the longitudinal components (E_z and H_z):

$$E_\varphi = \frac{i}{\kappa^2}\left(\frac{il\beta}{r}E_z + \omega\mu_0\frac{dH_z}{dr}\right) \tag{6.8}$$

$$E_r = \frac{i}{\kappa^2}\left(-\beta\frac{dE_z}{dr} + \omega\mu_0\frac{il}{r}H_z\right) \tag{6.9}$$

$$H_\varphi = \frac{i}{\kappa^2}\left(\frac{il\beta}{r}H_z - \omega\varepsilon\frac{dE_z}{dr}\right) \tag{6.10}$$

and

$$H_r = \frac{i}{\kappa^2}\left(-\beta\frac{dH_z}{dr} - \omega\varepsilon\frac{il}{r}E_z\right) \tag{6.11}$$

*The details have been worked out in Sec. 4.4. Here the φ dependence is of the form $\exp(-il\varphi)$.

where

$$\kappa^2 = n^2 k_0^2 - \beta^2$$
$$= k_0^2[K - U^2] = k_0^2[K_0 - U^2 - K_0 h(r)] \tag{6.12}$$

and $U = \beta/k_0$, ε_0 being the permittivity of free space and k_0 the free space wave number.

We next write the wave equations [see Eqs. (2.13) and (2.14)]

$$\nabla^2 \mathscr{E} + \nabla\left[\mathscr{E} \cdot \frac{1}{K}\nabla K\right] + \frac{K(r)\omega^2}{c^2}\mathscr{E} = 0 \tag{6.13}$$

and

$$\nabla^2 \mathscr{H} + \frac{1}{K}(\nabla K) \times (\nabla \times \mathscr{H}) + \frac{K(r)\omega^2}{c^2}\mathscr{H} = 0 \tag{6.14}$$

If we take the z component of the above equations and use Eq. (6.5) we get

$$\nabla^2 \mathscr{E}_z + \frac{\partial}{\partial z}\left[\frac{1}{K}\frac{dK}{dr}\mathscr{E}_r\right] + \frac{K(r)\omega^2}{c^2}\mathscr{E}_z = 0 \tag{6.15}$$

and

$$\nabla^2 \mathscr{H}_z + \frac{1}{K}\frac{dK}{dr}\left[\frac{\partial \mathscr{H}_r}{\partial z} - \frac{\partial \mathscr{H}_z}{\partial r}\right] + \frac{K(r)\omega^2}{c^2}\mathscr{H}_z = 0 \tag{6.16}$$

We now use Eqs. (6.6) and (6.7) to obtain

$$\frac{d^2 E_z}{dr^2} + \frac{1}{r}\frac{dE_z}{dr} - \frac{l^2}{r^2}E_z - \beta^2 E_z - \frac{i\beta}{K}\frac{dK}{dr}E_r + \frac{K(r)\omega^2}{c^2}E_z = 0 \tag{6.17}$$

and

$$\frac{d^2 H_z}{dr^2} + \frac{1}{r}\frac{dH_z}{dr} - \frac{l^2}{r^2}H_z - \beta^2 H_z - \frac{i\beta}{K}\frac{dK}{dr}H_r - \frac{1}{K}\frac{dK}{dr}\frac{dH_z}{dr} + \frac{K(r)\omega^2}{c^2}H_z = 0 \tag{6.18}$$

Now

$$-\frac{i\beta}{K}\frac{dK}{dr}E_r = -\frac{\beta}{K_0(1-h)}\frac{K_0 h'}{k_0^2[K_0 - U^2 - K_0 h(r)]}\left[-\beta\frac{dE_z}{dr} + \omega\mu_0\frac{il}{r}H_z\right] \tag{6.19}$$

and

$$-\frac{i\beta}{K}\frac{dK}{dr}H_r = \frac{\beta}{K_0(1-h)}\frac{K_0h'}{k_0^2[K_0-U^2-K_0h(r)]}$$

$$\times \left[\beta\frac{dH_z}{dr}+\omega\varepsilon_0K_0(1-h)\frac{il}{r}E_z\right] \quad (6.20)$$

On substitution we get

$$\phi''+\left[\frac{1}{r}+\frac{U^2h'(r)}{(1-h)(K_0-U^2-K_0h)}\right]\phi'+\left[k_0^2(K_0-U^2-K_0h)-\frac{l^2}{r^2}\right]\phi$$

$$=-\frac{lUK_0^{1/2}h'}{r(1-h)(K_0-U^2-K_0h)}\psi \quad (6.21)$$

and

$$\psi''+\left[\frac{1}{r}+\frac{K_0h'}{K_0-U^2-K_0h}\right]\psi'+\left[k_0^2(K_0-U^2-K_0h)-\frac{l^2}{r^2}\right]\psi$$

$$=-\frac{lUK_0^{1/2}h'}{r(K_0-U^2-K_0h)}\phi \quad (6.22)$$

where primes denote differentiation with respect to r and

$$\phi = K_0^{1/4}E_z \quad (6.23)$$

$$\psi = -iK_0^{-1/4}(\mu_0/\varepsilon_0)^{1/2}H_z \quad (6.24)$$

Equations (6.21) and (6.22) form a coupled set of equations. The equations get decoupled if $l = 0$ or $h' = 0$; the former case corresponds to no angular variation of the modes, whereas the latter case corresponds to constant permittivity. In either case both TE and TM modes exist, since then $\psi = 0$ does not imply $\phi = 0$, and vice versa. In general, one can eliminate one of the two field components and obtain a fourth-order differential equation for the other field component. The solution of the resulting fourth-order differential equation is extremely difficult to obtain. However, if the fields are appreciable only near the axis (i.e., near $r = 0$), where the dielectric constant is K_0, then β^2 will be very close to $k_0^2K_0$. Thus, if we define

$$\chi = 1-\frac{\beta^2}{k_0^2K_0} = 1-\frac{U^2}{K_0} \quad (6.25)$$

then for modes confined near the axis we must have

$$\chi \ll 1 \quad (6.26)$$

Substituting for U^2 from Eq. (6.25) in Eqs. (6.21) and (6.22) we get

$$\phi'' + \left[\frac{1}{r} + \frac{(1-\chi)h'}{(1-h)(\chi-h)}\right]\phi' + \left[k_0^2 K_0(\chi-h) - \frac{l^2}{r^2}\right]\phi = \frac{-l(1-\chi)^{1/2}h'}{r(1-h)(\chi-h)}\psi$$

(6.27)

and

$$\psi'' + \left[\frac{1}{r} + \frac{h'}{(\chi-h)}\right]\psi' + \left[k_0^2 K_0(\chi-h) - \frac{l^2}{r^2}\right]\psi = \frac{-l(1-\chi)^{1/2}h'}{r(\chi-h)}\phi$$

(6.28)

If we now assume Eqs. (6.4) and (6.26) to be valid, we would get

$$M\phi = \psi$$

(6.29)

and

$$M\psi = \phi$$

(6.30)

where

$$M = \frac{r(\chi-h)}{-lh'}\left\{\frac{d^2}{dr^2} + \left(\frac{1}{r} + \frac{h'}{\chi-h}\right)\frac{d}{dr} + \left[k_0^2 K_0(\chi-h) - \frac{l^2}{r^2}\right]\right\}$$

(6.31)

Thus

$$MM\phi = M\psi = \phi$$

(6.32)

and

$$MM\psi = M\phi = \psi$$

(6.33)

Consequently, the functions satisfying the equations

$$MG_j = +G_j \qquad j = 1, 3$$

(6.34)

$$MG_j = -G_j \qquad j = 2, 4$$

(6.35)

are also solutions of Eqs. (6.32) and (6.33).

Since the solutions of Eqs. (6.34) and (6.35) are linearly independent we may choose

$$\psi = \sum_{j=1}^{4} A_j G_j$$

(6.36)

Then

$$\phi = M\psi = \sum_{j=1}^{4} A_j M G_j \tag{6.37}$$

or

$$\phi = \sum_{j=1}^{4} (-1)^{j+1} A_j G_j \tag{6.38}$$

The choice for labeling the solutions of Eq. (6.34) by $j = 1, 3$ and the solutions of Eq. (6.35) by $j = 2, 4$ facilitates the writing of the solution for ϕ in a compact form.

In order to determine E_z and H_z, we must solve Eqs. (6.34) and (6.35). Substituting the expression for the operator in Eqs. (6.34) and (6.35) we obtain

$$\frac{d^2 G_j}{dr^2} + \left(\frac{1}{r} + \frac{h'}{\chi - h}\right)\frac{dG_j}{dr} + \left[b^2\left(1 - \frac{h}{\chi}\right) - \frac{l^2}{r^2} \pm \frac{lh'}{r(\chi - h)}\right]G_j(r) = 0 \tag{6.39}$$

where $b^2 = k_0^2 K_0 \chi$ and the upper and lower signs correspond to $j = 1, 3$ and $j = 2, 4$, respectively. Let G_3 and G_4 be the solutions of Eqs. (6.34) and (6.35), which are singular at the origin, then

$$E_z = K_0^{-1/4}\phi = K_0^{-1/4}(A_1 G_1 - A_2 G_2) \tag{6.40}$$

and

$$H_z = \frac{iK_0^{1/4}}{\eta_0}\psi = \frac{iK_0^{1/4}}{\eta_0}(A_1 G_1 + A_2 G_2) \tag{6.41}$$

where $\eta_0 = (\mu_0/\varepsilon_0)^{1/2}$.

Since $l = 0$, G_1, and G_2 are identical [see Eq. (6.39)] and we may choose $A_1 = A_2$ (giving rise to TE modes) or $A_1 = -A_2$ (giving rise to TM modes).

In order to obtain the transverse field components we may set either A_1 or A_2 equal to zero. Now, with $A_2 = 0$ and $A_1 = 1$, and using Eq. (6.8), we obtain

$$
\begin{aligned}
E_\varphi &= \frac{i}{\kappa^2}\left(\frac{il\beta}{r}K_0^{-1/4}G_1 + \omega\mu_0\frac{iK_0^{1/4}}{\eta_0}\frac{dG_1}{dr}\right) \\
&= -\frac{\beta K_0^{-1/4}}{k_0^2[K_0 - U^2 - K_0 h(r)]}\left[\frac{(\omega/\beta)(\varepsilon_0\mu_0)^{1/2}K_0^{1/2}r(dG_1/dr) + lG_1}{r}\right] \\
&= -\frac{(1-\chi)^{1/2}}{k_0 K_0^{3/4}(\chi - h)}\left[\frac{(1-\chi)^{-1/2}r(dG_1/dr) + lG_1}{r}\right]
\end{aligned} \tag{6.42}
$$

or

$$E_\varphi \approx -\Gamma\Phi_1 \tag{6.43}$$

where

$$\Gamma = \frac{1}{k_0 r_t K_0^{3/4} \chi} \tag{6.44}$$

and

$$\Phi_1(r) = \frac{r(dG_1/dr) + lG_1}{(r/r_t)[1 - (h/\chi)]} \tag{6.45}$$

where, in order to make Γ dimensionless, we have introduced the parameter r_t which is defined by the following equation

$$h(r_t) = \chi \tag{6.46}$$

Further, in writing Eq. (6.45) we have assumed $\chi \ll 1$. In a similar manner, we can calculate E_r and the transverse components of the magnetic field; the final results can be written in the following compact form

$$\mathbf{E}_t^{(j)} = (\mp i\hat{\mathbf{r}} - \hat{\boldsymbol{\varphi}})\Gamma\Phi_j(r) \qquad j = 1, 2 \tag{6.47a}$$

$$\mathbf{H}_t^{(j)} = \frac{K_0^{1/2}}{\eta_0}\hat{\mathbf{z}} \times \mathbf{E}_t^{(j)} \tag{6.47b}$$

where

$$\Phi_j(r) = \frac{r(dG_j/dr) \pm lG_j}{(r/r_t)[1 - (h/\chi)]} \qquad j = 1, 2 \tag{6.48}$$

$\hat{\mathbf{r}}$, $\hat{\boldsymbol{\varphi}}$, and $\hat{\mathbf{z}}$ denote unit vectors along the r, φ, and z directions, respectively, and the subscript t on $\mathbf{E}$ and $\mathbf{H}$ implies that we are considering the component transverse to the direction of propagation. Now, Eq. (6.48) can be rewritten as

$$\frac{r}{r_t}[1 - (h/\chi)]\Phi_j(r) = r\frac{dG_j}{dr} \pm lG_j \tag{6.49}$$

On differentiation, we get

$$\frac{1}{r_t}\left(1 - \frac{h}{\chi}\right)\Phi_j + \frac{r}{r_t}\left(-\frac{h'}{\chi}\right)\Phi_j + \left(\frac{r}{r_t}\right)\left(1 - \frac{h}{\chi}\right)\Phi_j' = rG_j'' + (1 \pm l)G_j' \tag{6.50}$$

If we substitute for G_j'' from Eq. (6.39) and then substitute for G_j' from Eq. (6.49) we obtain, after some simplification,

$$G_j = -\frac{r_t}{b^2}\left[\Phi_j' \mp \frac{(l \pm 1)}{r}\Phi_j\right] \qquad j = 1, 2 \qquad (6.51)$$

Further

$$G_j' = -\frac{r_t}{b^2}\left[\Phi_j'' \mp \frac{(l \pm 1)}{r}\Phi_j' \pm \frac{(l \pm 1)}{r^2}\Phi_j\right] \qquad j = 1, 2 \qquad (6.52)$$

If we substitute the above expressions for G_j and G_j' in Eq. (6.48) we would get

$$\Phi_j'' + \frac{1}{r}\Phi_j' + \left[b^2\left(1 - \frac{h}{\chi}\right) - \frac{(l \pm 1)^2}{r^2}\right]\Phi_j = 0 \qquad j = 1, 2 \qquad (6.53)$$

where $b^2 = k_0^2 K_0 \chi$. It is interesting to note that all the transverse components are proportional to a single scalar function, Φ_j, which satisfies a second-order differential equation; of course, one should remember that the entire analysis is based on weak inhomogeneity [i.e., Eq. (6.4)].

6.2.   Solutions for a Square-Law Medium

We will obtain specific solutions for a square-law medium where the dielectric constant variation is of the form (see Chapter 5 and Appendix B)

$$K = K_0 - K_2 r^2 \qquad (6.54)$$

The above dielectric constant distribution extends up to the guide wall, which is at $r = a$. However, as in Chapter 5, we carry out the analysis under the infinite medium approximation in which we assume Eq. (6.54) to be valid for all values of r. Comparing Eq. (6.54) with Eq. (6.5), we get

$$
\begin{aligned}
h(r) &= (K_2/K_0)r^2 \\
&= \delta(r/a)^2
\end{aligned}
\qquad (6.55)
$$

where we have introduced the dimensionless quantity $\delta \equiv (K_2/K_0)a^2$ and according to Eq. (6.4) $\delta \ll 1$. We define

$$\rho = \left(\frac{K_2}{K_0}\frac{1}{\chi}\right)^{1/2} r = \left(\frac{\delta}{\chi}\right)^{1/2}\left(\frac{r}{a}\right) \qquad (6.56)$$

In terms of ρ, Eqs. (6.39) and (6.53) become

$$\frac{d^2G_j}{d\rho^2}+\left(\frac{1}{\rho}+\frac{2\rho}{1-\rho^2}\right)\frac{dG_j}{d\rho}+\left[\alpha^2(1-\rho^2)-\frac{l^2}{\rho^2}\pm\frac{2l}{1-\rho^2}\right]G_j=0 \qquad (6.57)$$

and

$$\frac{d^2\Phi_j}{d\rho^2}+\frac{1}{\rho}\frac{d\Phi_j}{d\rho}+\left[\alpha^2(1-\rho^2)-\frac{(l\pm1)^2}{\rho^2}\right]\Phi_j=0 \qquad (6.58)$$

where

$$\alpha^2=b^2(\chi/K_2)=k_0^2K_0(\chi^2/K_2)=(k_0a)^2K_0(\chi^2/\delta) \qquad (6.59)$$

If we make the transformation

$$\Phi=\rho^s\psi_1(\rho)$$

where $s=l\pm1$, we would obtain

$$\frac{d^2\psi_1}{d\rho^2}+\frac{2s+1}{\rho}\frac{d\psi_1}{d\rho}+\alpha^2(1-\rho^2)\psi_1(\rho)=0 \qquad (6.60)$$

In terms of the new variable $\xi=\alpha\rho^2$, Eq. (6.60) becomes

$$\xi\frac{d^2\psi_1}{d\xi^2}+(s+1)\frac{d\psi_1}{d\xi}+\tfrac{1}{4}(\alpha-\xi)\psi_1(\xi)=0 \qquad (6.61)$$

If we substitute

$$\psi_1=e^{-\xi/2}L(\xi)$$

we would get

$$\xi\frac{d^2L}{d\xi^2}+(s+1-\xi)\frac{dL}{d\xi}+\tfrac{1}{4}(\alpha-2s-2)L(\xi)=0 \qquad (6.62)$$

which is the equation satisfied by associated Laguerre functions (see, for example, Irving and Mullineux, 1959, p. 99). The solution can be expressed in the form of an infinite power series which behaves as $\exp(\xi)$ for large values of ξ. However, if $\tfrac{1}{4}(\alpha-2s-2)$ is an integer, i.e.,

$$\alpha=4m+2l \qquad (6.63)$$

the series becomes a polynomial which is known as the associated Laguerre polynomial. Thus the solution of Eq. (6.58) can be written as

$$\Phi_j=\rho^{l\pm1}\exp(-\tfrac{1}{2}\alpha\rho^2)L^{(l\pm1)}_{(m-1/2\pm1/2)}(\alpha\rho^2) \qquad j=1,2 \qquad (6.64)$$

where

$$L_n^k(x) = \Gamma(n+k+1) \sum_{s=0}^{n} \frac{(-x)^s}{\Gamma(k+s+1)(n-s)!\,s!} \qquad (6.65)$$

which is a polynomial of degree n in x. Further, $m = 0, 1, 2, \ldots$ for $j = 1$, while $m = 1, 2, 3, \ldots$ for $j = 2$. For $l = 0$, $\Phi_1 = \Phi_2$, but the Φ_2 equation must be used since $L^{(-1)}$ is not defined. The $l = m = 0$ mode does not exist since it would be unbounded at $\rho = 0$. The functions G_j can be obtained by using Eq. (6.51) and are given by

$$G_j \sim \rho^l \exp[(-\alpha/2)\rho^2] N_l^{(j)}(\alpha, \rho^2) \qquad j = 1, 2 \qquad (6.66)$$

where

$$N_l^{(j)}(4m+2l, \rho^2) = \sum_{p=0}^{m} \frac{(-1)^p l!\,m!\,(4m+2l)^p}{(m-p)!\,p!\,(l+p)!}\left(1 - \frac{2p}{4m+l\pm l}\right)\rho^{2p} \qquad j = 1, 2$$

$$(6.67)$$

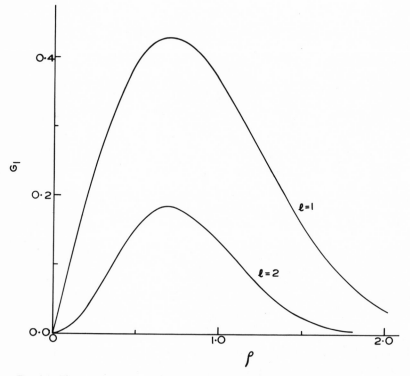

Fig. 6.1. The variation of G_1 with ρ for $m = 0, l = 1, 2$ (after Kurtz and Streifer, 1969a; reprinted by permission).

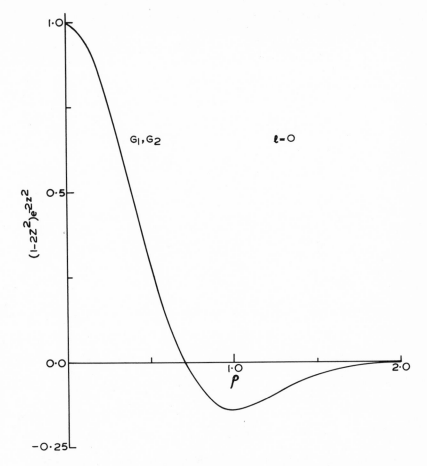

Fig. 6.2. The variation of G_1 and G_2 with ρ for $m = 1$, $l = 0$ (after Kurtz and Streifer, 1969a; reprinted by permission).

$m = 0, 1, 2, \ldots$ when $j = 1$ and $m = 1, 2, 3, \ldots$ when $j = 2$. The functions $N_l^{(j)}$ are polynomials, and explicit forms of these polynomials are given by Kurtz and Streifer (1969a). For example, for $m = 0$

$$G_1 = \rho^l \exp(-l\rho^2) \qquad (6.68)$$

Some of the G functions for the low-order modes are plotted in Figs. 6.1 and 6.2. The functions G_j satisfy the following orthogonality condition

$$\int_0^\infty G_j(4m + 2l, \rho) G_j(4m' + 2l, \rho)\rho \, d\rho = 0 \qquad \text{if } m \neq m' \quad (6.69)$$

The corresponding propagation constants can easily be determined. Using Eqs. (6.59) and (6.63) we obtain

$$\chi_m = \frac{(4m+2l)\delta^{1/2}}{k_0 a K_0^{1/2}} \qquad (6.70)$$

and since

$$\beta_m = k_0 K_0^{1/2}(1-\chi_m)^{1/2} \qquad (6.71)$$

we get

$$\beta_m \approx k_0 K_0^{1/2}\left[1 - \frac{(2m+l)\delta^{1/2}}{k_0 a K_0^{1/2}}\right] \qquad (6.72)$$

when the subscript m refers to the mode order. The theory developed will be valid when $\chi_m \ll 1$, which will indeed be the case when $k_0 a \gg 1$, and then β_m will be quite close to $k_0 K_0^{1/2}$. Further, the modes will be almost transverse when

$$\Gamma \gg 1$$

[see Eq. (6.46)]. Now

$$\Gamma = \frac{1}{k_0 r_t K_0^{3/4}\chi} = \frac{1}{k_0(\chi/\delta)^{1/2} a K_0^{3/4}\chi} \qquad (6.73)$$

where we have used the relation $h(r_t) = \chi$, which for a square-law medium gives $r_t = (\chi/\delta)^{1/2}a$. Substituting the value for χ, Eq. (6.73) becomes

$$\Gamma = \frac{(k_0 a)^{1/2}}{(4m+2l)^{3/2}\delta^{1/4}} \gg 1 \qquad (6.74)$$

Thus the modes are almost transverse for many modes when $k_0 a$ is large.

We notice that under the present approximation scheme the group velocity $(=d\beta/dk_0)$ is independent of the mode number [see Eq. (6.72)].

6.3. Asymptotic Solutions for Arbitrary Profiles

Kurtz and Streifer (1969b) have obtained asymptotic solutions (valid for large values of the dimensionless parameter br_t) for arbitrary, but weak inhomogeneity. The equation which determines the propagation constants is given by

$$b \int_0^{r_t} [1-(h/\chi)]^{1/2}\, dr = [m+(l/2)]\pi \qquad (6.75)$$

where, as before,

$$h(r_t) = \chi$$

Thus

$$B \int_0^1 [1 - (h/\chi)]^{1/2} \, dz = [m + (l/2)]\pi \tag{6.76}$$

where,

$$B = br_t = k_0 r_t (K_0 \chi)^{1/2} \tag{6.77}$$

For a power law profile

$$h(r) = \delta (r/a)^p \qquad p = 1, 2, \ldots \tag{6.78}$$

Thus

$$\delta (r_t/a)^p = \chi$$

and

$$h(z) = \chi z^p \qquad z = \frac{r}{r_t} \tag{6.79}$$

Substituting for $h(z)$ from Eq. (6.79) in Eq. (6.75) we obtain

$$B \int_0^1 (1 - z^p)^{1/2} \, dz = (4m + 2l)(\pi/4) \tag{6.80}$$

The integral can be evaluated in terms of gamma functions and one gets

$$B = (4m + 2l)H(p) \tag{6.81}$$

where

$$H(p) = \frac{\pi}{4} \frac{\Gamma[\frac{3}{2} + (1/p)]}{\Gamma[1 + (1/p)]\Gamma(\frac{3}{2})} \tag{6.82}$$

The function $H(p)$ is a monotonically decreasing function of p and

$$H(1) = \frac{3}{2} \frac{\pi}{4}$$

$$H(\infty) = \frac{\pi}{4}$$

As mentioned earlier, the asymptotic solutions are valid when $B \gg 1$ which is valid for a large number of modes. The corresponding propagation constants can be obtained by using Eq. (6.77) and are given by

$$\chi = 1 - \frac{\beta^2}{k_0^2 K_0} = \left[\frac{(4m + 2l)H(p)}{k_0 a K_0^{1/2}}\right]^{2p/(p+2)} \delta^{2/(p+2)} \tag{6.83}$$

For a square-law medium $p = 2$ and since $H(2) = 1$ one obtains the same expression for χ as obtained by the rigorous analysis [cf. Eq. (6.70)].

The corresponding field patterns can be expressed in terms of the Bessel functions. For example, the asymptotic expression for G_j is given by

$$G_j \sim (-1)^{m+1} \left(\frac{2}{z}\right)^{1/2} \left(1 - \frac{h}{\chi}\right)^{1/4} \left(\frac{3}{2}B\Lambda\right)^{-1/6} Ai'\left[-\left(\frac{3}{2}B\Lambda\right)^{2/3}\right]$$

(6.84)

where

$$\Lambda = i \int_1^z [(h/\chi) - 1]^{1/2} \, dz \qquad \text{for } 1 \leqslant z$$

$$= \int_z^1 [1 - (h/\chi)]^{1/2} \, dz \qquad \text{for } 1 > z$$

and the prime denotes the derivative of the Airy function, $Ai(x)$, with respect to the argument. Further

$$J_{\pm 2/3}(\xi) = (\sqrt{3}/2x)[\pm\sqrt{3}Ai'(-x) + Bi'(-x)] \qquad x = (\tfrac{3}{2}\xi)^{2/3}$$

Thus

$$Ai'(-x) = (x/3)[J_{2/3}(\xi) - J_{-2/3}(\xi)]$$

(6.85)

G_1 and G_2 are the same in the approximation under which the asymptotic solutions have been derived. Explicit z dependence of G_j [as obtained from Eq. (6.84) for $p = 2$] agrees quite well with the exact expressions except near $z = 0$ (see Kurtz and Streifer, 1969b).

Equation (6.80) can be readily integrated for the Pöschl–Teller profile and for the Epstein-type profile. For the Pöschl–Teller profile

$$h(r) = \delta_{PT} \tan^2[(\pi/2)(r/a_{PT})]$$

(6.86)

and Eq. (6.80) gives

$$\chi = (2\gamma_{PT} + \gamma_{PT}^2)\delta_{PT}$$

(6.87)

where

$$\gamma_{PT} = \frac{\Gamma(4m + 2l)}{4k_0 a_{PT} K_0^{1/2} \delta_{PT}^{1/2}}$$

(6.88)

Similarly for the Epstein profile

$$h(r) = \delta_E \tanh^2(r/a_E)$$

(6.89)

and Eq. (6.80) gives

$$\chi = (2\gamma_E - \gamma_E^2)\delta_E \tag{6.90}$$

where

$$\gamma_E = \frac{2m + l}{k_0 a_E (K_0 \delta_E)^{1/2}} \tag{6.91}$$

From Eqs. (6.87) and (6.90) one can find the propagation constants by using Eq. (6.25). It should be noted that in the first case if δ_{PT} and a_{PT} tend to infinity such that δ_{PT}/a_{PT}^2 tends to a finite value one obtains the quadratic profile and the results obtained agree with the rigorous analysis; similarly for the Epstein profile.

6.4. Approximate Methods

The wave equation can be solved exactly only for very few dielectric constant profiles. Approximate methods are therefore expected to play an important role in most inhomogeneous waveguides. We will discuss three important approximate methods, that is, the perturbation method, the WKB method, and the variational method.

6.5. The Perturbation Method

In this section we have used the first-order perturbation theory to calculate the propagation constants corresponding to systems for which exact solutions are not possible. The perturbation theory technique is discussed in any book on quantum mechanics (see, for example, Ghatak and Lokanathan, 1975) and has been used by many workers (Gambling and Matsumura, 1973; Kitano *et al.*, 1973; Kumar *et al.*, 1974; Marcuse, 1973a; Thyagarajan and Ghatak, 1974; Yamada and Inabe, 1974). The method consists of finding the eigenvalues of a Hermitian* operator H where the operator H is such that it can be written as a sum of two parts: one of these parts, H_0, is such that its eigenvalue spectrum and the corresponding eigenfunctions are fully known; and the other part H' can be considered as a small perturbation to H_0. Thus, we wish to solve the eigenvalue equation

$$H\psi = W\psi \tag{6.92}$$

*Hermitian operators are defined in any book on mathematical physics (see, for example, Margenau and Murphy, 1956). Hermitian operators are characterized by real eigenvalues and their eigenfunctions form an orthonormal set.

where

$$H = H_0 + H' \tag{6.93}$$

and

$$H_0 u_n = \lambda_n u_n \tag{6.94}$$

u_n being the *known* orthonormal eigenfunctions of H_0; λ_n being the corresponding eigenvalues. According to the first-order perturbation theory the eigenvalues and the eigenfunctions of the mth state are given by

$$W_m = \lambda_m + H'_{mm} \tag{6.95}$$

and

$$\psi_m = u_m + \sum_{\substack{m \\ m \neq n}} \frac{|H'_{nm}|^2}{E_m - E_n} \tag{6.96}$$

where

$$H'_{nm} = \int u_n^* H' u_m \, d\tau \tag{6.97}$$

The above relations are valid for states which are nondegenerate; i.e., when there is only one eigenfunction corresponding to an eigenvalue. For degenerate states the considerations are more involved.

We shall now use the perturbation method to study the propagation constants corresponding to a variety of interesting situations.

6.5.1. Perturbation Theory for Cladded Square-Law Media

We first consider a cladded square-law medium in which the dielectric constant distribution, $K(x)$, would be given by (see Fig. 6.3)

$$\begin{aligned} K(x) &= K_1 - \Delta(x^2/a^2) \qquad &|x| \leqslant a \\ &= K_2 \qquad &|x| > a \end{aligned} \tag{6.98}$$

We will carry out an analysis of the TE modes of such a waveguide. The only nonvanishing field components would be H_z, H_x, and E_y; and E_y would satisfy (see Sec. 2.3)

$$\frac{d^2\psi}{dx^2} + [K(x)k_0^2 - \beta^2]\psi = 0 \tag{6.99}$$

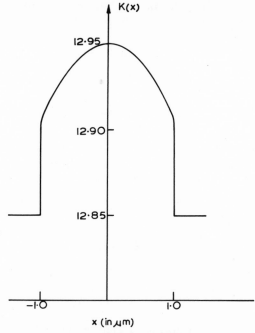

Fig. 6.3. Typical variation of the dielectric constant in a cladded square-law medium.

which may be rewritten in the form

$$\frac{d^2\psi}{dx^2}+\{[K^{(0)}(x)+K'(x)]k_0^2-\beta^2\}\psi=0 \qquad (6.100)$$

where

$$K^{(0)}(x)=K_1-\Delta(x^2/a^2) \qquad (6.101)$$

and $K'(x)$, which is to be considered as the perturbation, is given by

$$\begin{aligned} K'(x) &= 0 & |x| &\leqslant a \\ &= K_2-K^{(0)}(x) & |x| &> a \end{aligned} \qquad (6.102)$$

The modes and the propagation constants for the unperturbed equation (i.e., with $K'=0$) have been derived in Chapter 3 and are given by

$$\psi_n=\frac{1}{(2^n n!\alpha\sqrt{\pi})^{1/2}}H_n(\xi)\exp(-\tfrac{1}{2}\xi^2)$$

$$\beta_n^{(0)}=[K_1k_0^2-k_0(\sqrt{\Delta}/a)(2n+1)]^{1/2} \qquad n=0,1,2,\dots \qquad (6.103)$$

where $\xi = x/\alpha$ and $\alpha^2 = (a/k_0\sqrt{\Delta})$. Thus, in first-order perturbation theory the propagation constants would be given by

$$\beta_n^2 = [\beta_n^{(0)}]^2 + [\beta_n^{(1)}]^2 \qquad (6.104)$$

where

$$[\beta_n^{(1)}]^2 = k_0^2 \int_{-\infty}^{+\infty} \psi_n^*(x) K'(x) \psi_n(x)\, dx \qquad (6.105)$$

If we substitute the Hermite–Gauss functions for $\psi_n(x)$, then the integral in Eq. (6.105) can always be put in terms of the error function; e.g., for $n = 0$

$$[\beta_0^{(1)}]^2 = k_0^2\{[(\Delta/2\xi_0^2) - g^2]\mathrm{erfc}(\xi_0) + (\Delta/\sqrt{\pi}\xi_0)\,e^{-\xi_0^2}\} \qquad (6.106)$$

where

$$\xi_0 = a/\alpha$$
$$g^2 = K_1 - K_2$$

and

$$\mathrm{erfc}(\xi_0) = 1 - (2/\sqrt{\pi}) \int_0^{\xi_0} e^{-t^2}\, dt$$

The corresponding group velocity can easily be calculated. In particular, for $K_1 = 12.95$, $K_2 = 12.85$, $\Delta = 0.044$, $a = 1\ \mu\mathrm{m}$, and $\lambda = 0.8383\ \mu\mathrm{m}$ (see Fig. 6.3) [which represent typical values for a $p - n$ junction (Zachos and Ripper, 1969)], the group velocity for the lowest order mode is $0.2004c$ whereas the corresponding value for an infinitely extended square-law medium is $0.196c$. Thus the perturbation theory predicts a 2.2% increase in the group velocity of the mode.

The above choice of $K^{(0)}(x)$ and $K'(x)$ are expected to give good results for low order modes; for modes near cutoff we must choose

$$\begin{aligned} K^{(0)}(x) &= K_1 \qquad |x| \le a \\ &= K_2 \qquad |x| > a \end{aligned} \qquad (6.107)$$

and

$$\begin{aligned} K'(x) &= -\Delta(x^2/a^2) \qquad |x| \le a \\ &= 0 \qquad |x| > a \end{aligned} \qquad (6.108)$$

The modal analysis for the unperturbed equation has been carried out in Sec. 2.3 and the normalized wave functions for *symmetric* TE modes are given by

$$\psi = \frac{1}{[a+(1/\kappa)]^{1/2}} \cos px \qquad\qquad |x| \leq a$$

$$= \frac{1}{[a+(1/\kappa)]^{1/2}} \cos pa \, \exp[-\kappa(|x|-a)] \qquad |x| \geq a$$

(6.109)

where $p^2 = (\omega^2/c^2)K_1 - \beta^2$ and $\kappa^2 = \beta^2 - (\omega^2/c^2)K_2$. The unperturbed propagation constants, $\beta^{(0)}$, are roots of the equation

$$p \tan pa = \kappa \qquad\qquad (6.110)$$

Thus, if we write

$$\beta^2 = \beta^{(0)2} + \beta^{(1)2} \qquad\qquad (6.111)$$

then the perturbation $\beta^{(1)2}$ would be given by

$$\beta^{(1)2} = -k_0^2 \int_{-a}^{+a} \psi^* \Delta(x^2/a^2)\psi \, dx$$

$$= -ak_0^2 \frac{\Delta}{[a+(1/\kappa)]} \left\{ \frac{1}{3} \mp \left[\frac{\cos 2pa}{2p^2a^2} + \left(1 - \frac{1}{2p^2a^2}\right) \frac{\sin 2pa}{2pa} \right] \right\} \qquad (6.112)$$

where the upper and lower signs correspond to antisymmetric and symmetric modes. The analysis is expected to give accurate results for modes near cutoff. Numerical calculations have been carried out by Kumar *et al.* (1974).

6.5.2. Perturbation Theory for Planar Waveguides Formed by the Outdiffusion Technique

Some of the planar waveguides fabricated by Kaminow and Carruthers (1973) using the out-diffusion technique have the following dielectric constant profile (see Sec. A.3)

$$K(x) = K_0 \qquad\qquad x < 0$$

$$= K_1 + \Delta K \, \text{erfc}(2x/\pi a) \qquad x > 0$$

(6.113)

Numerical calculations for such profiles have been reported by Smithgall and Dabby (1973). Since the above dielectric constant distribution resembles the exponential distribution, one can consider it to be a perturbed form of the following distribution (see Sec. 3.2)

$$K(x) = K_0 \qquad\qquad x < 0$$

$$= K_1 + \Delta K \, \exp(-x/d) \qquad x > 0$$

(6.114)

The modes for such a dielectric constant distribution have been derived in Chapter 3; they are Bessel functions in the region $x > 0$ and exponentially decaying functions in the region $x < 0$ [see Eqs. (3.48) and (3.49)]. Using these as the unperturbed wave functions and assuming the perturbation to be zero for $x < 0$ and $\Delta K[\mathrm{erfc}(2x/\pi a) - \exp(-x/d)]$ for $x > 0$, the propagation constants for the waveguide described by Eq. (6.113) can be easily calculated (Kumar et al., 1974). Typically, for $K_0 = 2.268$, $\Delta K = 0.9185$, $a = 0.56\ \mu$, $d = 0.4767\ \mu$, and $k_0 = 12\ \mu^{-1}$ the unperturbed eigenvalue, $\beta^{(0)}$, for the lowest order mode obtained by solving Eq. (3.50) comes out to be $1.6425\ k_0$. Applying the first-order perturbation theory the value changes to $1.6649\ k_0$ which is within 0.08% of the exact value obtained through a numerical solution of the differential equation (Smithgall and Dabby, 1973).

6.5.3. Perturbation Theory for Taking into Account Higher Order Terms in a SELFOC Fiber

Kitano et al. (1973) have used the perturbation theory to calculate the propagation constants (using the scalar wave equation) for a SELFOC fiber characterized by the following dielectric constant variation

$$K(x, y) = K^{(0)} + K' \tag{6.115}$$

where

$$K^{(0)} = K_0 - K_2 x^2 - K_3 y^2 \tag{6.116}$$

and

$$K' = K_0(K_{4x} x^4 + K_{4y} y^4 + K_{4xy} x^2 y^2) \tag{6.117}$$

The modal analysis for the unperturbed part has already been carried out; the wave functions are Hermite–Gauss functions [see Eq. (5.15)]. Assuming K' to be a perturbation, the first-order perturbation theory gives (Kitano et al., 1973)

$$
\begin{aligned}
\beta = {} & \frac{\omega}{c} K_0^{1/2} - \left[(m + \tfrac{1}{2})\left(\frac{K_2}{K_0}\right)^{1/2} + (n + \tfrac{1}{2})\left(\frac{K_3}{K_0}\right)^{1/2} \right] \\
& + \frac{1}{8}\frac{c}{\omega} K_0^{1/2}\Bigg[3(2m^2 + 2m + 1)K_{4x}\frac{K_0}{K_2} \\
& + 3(2n^2 + 2n + 1)K_{4y}\frac{K_0}{K_3} - 4(m + \tfrac{1}{2})(n + \tfrac{1}{2})K_{4xy}\frac{K_0}{(K_2 K_3)^{1/2}} \\
& - (2m + 1)^2\frac{K_2}{K_0} - (2n + 1)^2\frac{K_3}{K_0} \\
& - 2(2m + 1)(2n + 1)\frac{(K_2 K_3)^{1/2}}{K_0} \Bigg]
\end{aligned} \tag{6.118}
$$

where m and n are the mode numbers. A perturbation theory analysis, taking into account quartic terms in the dielectric constant variation, has also been carried out by Gambling and Matsumura (1973) for cylindrically symmetric modes.

6.5.4. Perturbation Theory for Studying the Effect of ∇K Term in Lens-like Media

For a general inhomogeneous medium with dielectric constant variation $K(x, y)$, the x and y component of the electric field would satisfy* [see Eq. (2.13)]

$$(\nabla_t^2 + \kappa^2)E_x + \frac{\partial}{\partial x}\left(\frac{1}{K}\frac{\partial K}{\partial x}E_x + \frac{1}{K}\frac{\partial K}{\partial y}E_y\right) = \beta^2 E_x \qquad (6.119)$$

$$(\nabla_t^2 + \kappa^2)E_y + \frac{\partial}{\partial y}\left(\frac{1}{K}\frac{\partial K}{\partial x}E_x + \frac{1}{K}\frac{\partial K}{\partial y}E_y\right) = \beta^2 E_y \qquad (6.120)$$

where $\Delta_t^2 = (\partial^2/\partial x^2) + (\partial^2/\partial y^2)$ and $\kappa^2 = \omega^2\mu K(x, y)$. The time and z variation has been assumed to be of the form $\exp[i(\omega t - \beta z)]$. Equations (6.119) and (6.120) can be written in the form

$$P\psi = \beta^2\psi \qquad (6.121)$$

where

$$\psi = \begin{pmatrix} E_x \\ E_y \end{pmatrix}$$

$$P = P_0 + P'$$

$$P_0 = \begin{pmatrix} L & 0 \\ 0 & L \end{pmatrix}$$

$$L \equiv \nabla_t^2 + \kappa^2$$

$$P' = \begin{vmatrix} \dfrac{\partial}{\partial x}\left(\dfrac{1}{K}\dfrac{\partial K}{\partial x}\right) & \dfrac{\partial}{\partial x}\left(\dfrac{1}{K}\dfrac{\partial K}{\partial y}\right) \\ \dfrac{\partial}{\partial y}\left(\dfrac{1}{K}\dfrac{\partial K}{\partial x}\right) & \dfrac{\partial}{\partial y}\left(\dfrac{1}{K}\dfrac{\partial K}{\partial y}\right) \end{vmatrix} \qquad (6.122)$$

P' being considered as a perturbation to the operation P_0. Let the solution of the unperturbed equation

$$P_0\psi = \beta^2\psi \qquad (6.123)$$

*The treatment is based on a paper by Thyagarajan and Ghatak (1974).

be represented by a known function H_{pq}, where p and q represent mode numbers. For example, if we consider the case of a parabolic index medium characterized by

$$K(x, y) = K_0 - K_2 x^2 - K_3 y^2 \qquad (6.124)$$

where, in general, $K_2 \neq K_3$, the functions H_{pq} would be Hermite–Gauss functions [see Eq. (5.15)] and

$$\beta_{pq}^2 = \frac{\omega^2 K_0}{c^2} \left[1 - (2p+1) \frac{c}{\omega K_0} K_2^{1/2} - (2q+1) \frac{c}{\omega K_0} K_3^{1/2} \right]$$
$$p, q = 0, 1, 2, \ldots \qquad (6.125)$$

The modes represented by H_{pq} are, in general, nondegenerate with respect to p and q. But when $K_2 = K_3$, the H_{pq} are $(p+q+1)$-fold degenerate. We will first assume that K_2 and K_3 are such that the modes are nondegenerate. However, even then we will have two wave functions

$$\psi_1 = \begin{pmatrix} H_{pq} \\ 0 \end{pmatrix}$$
$$\psi_2 = \begin{pmatrix} 0 \\ H_{pq} \end{pmatrix} \qquad (6.126)$$

which would satisfy the unperturbed equations, giving the same eigenvalue β_{pq}^2. In order to study the effect of the ∇K term, the wave function is assumed to be of the form

$$\psi = C_1 \psi_1 + C_2 \psi_2$$

where the constants C_1 and C_2, and the perturbation $W_1 (= \beta^{(1)^2})$ will be determined by the equations

$$C_1(P'_{11} - W_1) + C_2 P'_{12} = 0$$
$$C_1 P'_{21} + C_2(P'_{22} - W_1) = 0 \qquad (6.127)$$

where

$$P'_{11} = \langle \psi_1 | P' | \psi_1 \rangle = \int_{-\infty}^{\infty} \int_{-\infty}^{\infty} dx\, dy\, (H_{pq}\ 0) \begin{pmatrix} O_1 & O_2 \\ O_3 & O_4 \end{pmatrix} \begin{pmatrix} H_{pq} \\ 0 \end{pmatrix} \qquad (6.128)$$

and so on, where the operators O_1, O_2, O_3, and O_4 are defined by

$$O_1 = \frac{\partial}{\partial x}\left(\frac{1}{K}\frac{\partial K}{\partial x}\right) \qquad O_2 = \frac{\partial}{\partial x}\left(\frac{1}{K}\frac{\partial K}{\partial y}\right)$$

$$O_3 = \frac{\partial}{\partial y}\left(\frac{1}{K}\frac{\partial K}{\partial x}\right) \qquad O_4 = \frac{\partial}{\partial y}\left(\frac{1}{K}\frac{\partial K}{\partial y}\right) \qquad (6.129)$$

Parabolic Index Medium with $K_2 \neq K_3$

For K_2, $K_3 \ll K_0$ we can write

$$P' = -\frac{2}{K_0}\begin{vmatrix} K_2\dfrac{\partial}{\partial x}(x & K_3 y\dfrac{\partial}{\partial x} \\ K_2 x\dfrac{\partial}{\partial y} & K_3\dfrac{\partial}{\partial y}(y \end{vmatrix} \qquad (6.130)$$

The matrix elements can be evaluated easily and we get the following two equations

$$C_1\left(W_1 + \frac{K_2}{K_0}\right) = 0$$

$$C_2\left(W_1 + \frac{K_3}{K_0}\right) = 0$$

Thus we have (i) $W_1 = -(K_2/K_0)$, $C_1 \neq 0$ and $C_2 = 0$, i.e., $\beta^2 = \beta_{pq}^2 - (K_2/K_0)$, $E_x \neq 0$ and $E_y = 0$; (ii) $W_1 = -(K_3/K_0)$, $C_2 \neq 0$ and $C_1 = 0$, i.e., $\beta^2 = \beta_{pq}^2 - (K_3/K_0)$, $E_x = 0$ and $E_y \neq 0$. The fundamental modes are such that only one transverse component is present. The longitudinal components are also present, and are small but finite. The analysis should find applications in the propagation of electromagnetic waves through *p–n* junctions (Zachos and Ripper, 1969).

Parabolic Index Medium with $K_2 = K_3$

It is to be noted that for this case one cannot carry over the perturbation theory developed in the last section and take the limit $K_2 \to K_3$. For such a case the theory will have to be developed *ab initio*, taking into consideration all of the degenerate modes. The details have been given by Thyagarajan and Ghatak (1974); for example, for $p + q = 2$ the following set of modes are obtained:

(i) $W_1 = -(K_2/K_0)$, i.e., $\beta^2 = \beta_{pq}^2 - (K_2/K_0)$ and the wave function can be either (a) $E_x = H_{20} + H_{02}$, $E_y = 0$ or (b) $E_x = 0$, $E_y = H_{20} + H_{02}$ or any linear combination of them.

(ii) $W_1 = -(3K_2/K_0)$, i.e., $\beta^2 = \beta_{pq}^2 - (3K_2/K_0)$ and the wave function can be either (a) $E_x = H_{20} - H_{02}$, $E_y = -\sqrt{2}H_{11}$ or (b) $E_x = \sqrt{2}H_{11}$, $E_y = H_{20} - H_{02}$ or any linear combination of them.

(iii) $W_1 = +(K_2/K_0)$, i.e., $\beta^2 = \beta_{pq}^2 + (K_2/K_0)$ and the wave function can be (a) $E_x = H_{20} - H_{02}$, $E_y = \sqrt{2}H_{11}$ or (b) $E_x = -\sqrt{2}H_{11}$, $E_y = H_{20} - H_{02}$ or any linear combination of them.

The above analysis should find applications in the study of electromagnetic wave propagation through SELFOC fibers.

6.6. The WKB Method

The WKB method has been used by many workers to study the propagation characteristics of optical waveguides (see, for example, Gloge and Marcatili, 1973; Gedeon, 1974; Marcuse, 1973b). For a continuous dielectric constant variation (in a slab waveguide) the propagation constants are determined from the relation (see, e.g., Schiff, 1968).

$$\int_{x_1}^{x_2} [K(x)k_0^2 - \beta_n^2]^{1/2}\, dx = (n + \tfrac{1}{2})\pi \qquad n = 0, 1, 2, \ldots \quad (6.131)$$

where x_1 and x_2 are the two turning points, i.e.,

$$K(x_i)k_0^2 = \beta_n^2 \qquad i = 1, 2 \qquad (6.132)$$

For a parabolic index distribution $K = K_0 - K_2 x^2$ and

$$x_1 = -\left(\frac{K_0 - \beta_n^2/k_0^2}{K_2}\right)^{1/2} \qquad x_2 = +\left(\frac{K_0 - \beta_n^2/k_0^2}{K_2}\right)^{1/2}$$

Thus

$$\int_{x_1}^{x_2} [K(x)k_0^2 - \beta_n^2]^{1/2}\, dx = \int_{-\left[\frac{K_0 - (\beta_n^2/k_0^2)}{K_2}\right]^{1/2}}^{+\left[\frac{K_0 - (\beta_n^2/k_0^2)}{K_2}\right]^{1/2}} [(K_0 k_0^2 - \beta_n^2) - K_2 k^2 x^2]^{1/2}\, dx$$

$$= \frac{K_0 k_0^2 - \beta_n^2}{k_0 K_2^{1/2}} \int_{-1}^{+1} (1 - t^2)^{1/2}\, dt$$

$$= \frac{\pi(K_0 k_0^2 - \beta_n^2)}{2k_0 K_2^{1/2}}$$

Substituting this in Eq. (6.131) we obtain

$$\beta_n^2 = K_0 k_0^2 - (2n+1)k_0 K_2^{1/2} \qquad (6.133)$$

which agrees with the exact solution* [see Eq. (3.17)]. Even for the dielectric constant variation of the form [see Eq. (3.66)]

$$K(x) = K_0 + \frac{K_2 - K_0}{\cosh^2 x/a} \qquad (6.134)$$

the WKB approximation gives accurate results. Indeed, on evaluating the integral in Eq. (6.131) one obtains

$$\beta_n^2 = K_0 k_0^2 + [k_0(K_2 - K_0)^{1/2} - (n + \tfrac{1}{2})(1/a)]^2 \qquad (6.135)$$

which agrees very well with the exact solution

$$\beta_n^2 = K_0 k_0^2 + \{(1/a)[k_0^2 a^2 (K_2 - K_0) + \tfrac{1}{4}]^{1/2} - (n + \tfrac{1}{2})(1/a)\} \qquad (6.136)$$

when $k_0^2 a^2 (K_2 - K_0) \gg \tfrac{1}{4}$.

For a discontinuous dielectric constant variation of the form

$$\begin{aligned} K(x) &= 1 & x < 0 \\ &= K_g(x) & x > 0 \end{aligned} \qquad (6.137)$$

the equation determining the propagation constants is modified to the following (see, for example, Gedeon, 1974):

$$\int_{x_1}^{x_2} [k(x)k_0^2 - \beta_n^2]^{1/2} \, dx = (n + \tfrac{1}{2})\pi + \phi_n \qquad n = 0, 1, 2, \ldots \qquad (6.138)$$

where

$$\begin{aligned} \phi_n &= 0 & \text{if } x_1 > 0 \\ &= \tan^{-1}\left[\frac{(\beta_n^2/k_0^2) - 1}{K_g(0) - (\beta_n^2/k^2)}\right] - \frac{\pi}{4} & \text{if } x_1 = 0 \end{aligned} \qquad (6.139)$$

Gedeon (1974) has studied the following three dielectric constant profiles

$$K_g(x) = n_g^2(x) = n_s^2 + 2n_s \Delta \exp(-x/a) \qquad (6.140)$$

$$= [n_s + \Delta \exp(-x/a)]^2 \qquad (6.141)$$

$$= \left\{ n_s + \Delta \exp\left[-\frac{(x-d)^2}{a^2}\right] \right\} \qquad (6.142)$$

---

*This is the well-known result in quantum mechanics where the WKB approximation is known to be exact for the linear harmonic oscillator.

Such dielectric constant profiles have indeed been obtained (see Appendix A). The WKB results for the propagation constants compare very well with the exact numerical results (see Table 6.1). The corresponding field patterns, between the turning points, are given by

$$E^{(n)}(x) \sim \frac{1}{[K(x)k_0^2 - \beta_n^2]^{1/4}} \cos\left[\int_{x_1}^{x} [K(x)k_0^2 - \beta_n^2]^{1/2}\, dx + \frac{\pi}{4} - \phi_n\right] \quad (6.143)$$

and, except near the turning points, the field patterns agree well with the exact results (Gedeon, 1974).

Marcuse (1973b) has carried out a WKB analysis of piecewise dielectric constant variation of the form

$$\begin{aligned}
K(x) &= 1 & -\infty < x < 0 \\[6pt]
&= K_0 + (K_1 - K_0)\frac{x}{x_1} & 0 < x < x_1 \\[6pt]
&= K_1 + (K_2 - K_1)\frac{x - x_1}{x_2 - x_1} & x_1 < x < x_2 \\[6pt]
&= K_2 & x_2 < x < \infty
\end{aligned} \quad (6.144)$$

For the above dielectric constant variation the wave equation for the TE modes [Eq. (2.44)] can be exactly solved; the results can be expressed in terms of Bessel functions in the regions $0 < x < x_1$ and $x_1 < x < x_2$, and

Table 6.1. Comparison between the Values of the Propagation Constants[a]

	$n_s = 2.177 \quad \Delta = 0.09837$ $K_g(x) = n^2(x)$ $= n_s^2 + 2n_s\Delta e^{-x/a}$		$a = 2.22726\,\mu\text{m}$ $K_g(x) = n^2(x)$ $= (n_s + \Delta e^{-x/a})^2$		$n_s = 1.512 \quad \Delta = 0.0833$ $d = 2.60\,\mu\text{m} \quad a = 2.66\,\mu\text{m}$ $n^2(x) = [n_s$ $+ \Delta \exp^{[-(x-d)^2/a^2]}]^2$	
n	Exact	WKB	Exact	WKB	Exact	WKB
0	2.24135	2.24160	2.24275	2.24290	1.58935	1.58925
1	2.22070	2.22075	2.22160	2.22160	1.57775	1.57765
2	2.20680	2.20680	2.20745	2.20740	1.56680	1.56670
3	2.19675	2.19675	1.19725	2.19715	1.55635	1.55650
4	2.18940	2.18940	2.18980	2.18970	1.54620	1.54705
5	2.18415	2.18415	2.18440	2.18430	1.53625	1.53580
6	2.18050	2.18050	2.18070	2.18060	1.52680	1.52655
7	2.17825	2.17825	2.17835	2.17830	1.51850	1.51830
8	2.17715	2.17715	2.17720	2.17715	—	—

[a] (β/k) obtained from rigorous theory and from WKB-analysis for graded-index waveguides with different index profiles of physical interest (after Gedeon, 1974).

exponentially decaying solutions in the regions $-\infty < x < 0$ and $x_2 < x < \infty$. The continuity of the fields at the interfaces give the transcendental equation which determines the propagation constants; however, the numerical calculations are, in general, extremely difficult. Using the WKB method, Marcuse (1973b) has given useful approximate solutions for the dielectric constant variation given by Eq. (6.144) and has obtained a closed form expression for the total number of modes that can be supported by the waveguide.

Gloge and Marcatali (1973) have used the WKB method to study the modes for graded core fibers and have obtained the following expression for the total number of modes

$$m(\beta) = \int_0^R [k_0^2 K(r) - \beta^2]^{1/2} r \, dr \qquad (6.145)$$

where $m(\beta)$ denotes the number of modes having a propagation constant larger than β, and the upper limit corresponds to the second turning point. Assuming

$$
\begin{aligned}
K(r) &= K_0[1 - 2\Delta(r/a)^\alpha] & r < a \\
&= K_0[1 - 2\Delta] & r > a
\end{aligned}
\qquad (6.146)
$$

one obtains

$$m(\beta) = a^2 \Delta k_0 K_0 \frac{\alpha}{\alpha + 2} \left(\frac{k_0^2 K_0 - \beta^2}{2\Delta k^2 K_0} \right)^{(2/\alpha)+1} \qquad (6.147)$$

Notice that $\alpha = 2$ corresponds to a SELFOC fiber.

6.7. The Variational Method

Assuming a z dependence of the field to be of the form $\exp(-i\beta z)$, the scalar wave equation becomes

$$\nabla_t^2 \psi + [k_0^2 K(x, y) - \beta^2]\psi = 0 \qquad (6.148)$$

where $\nabla_t^2 \equiv \nabla^2 - (\partial^2/\partial z^2)$. The variational formula for β^2 is given by

$$\beta^2 = \frac{\{\int [k_0^2 \phi^2 - (\nabla \phi)^2] \, ds\}}{\int \phi^2 \, ds} \qquad (6.149)$$

where the surface integral has to be carried out over the whole transverse surface and the trial function ϕ must be continuous on the entire integral surface. The trial functions ϕ contain a few parameters C_i and one chooses these parameters in a way such that

$$\frac{\partial \beta^2}{\partial C_i} = 0 \tag{6.150}$$

Using linear combinations of parabolic cylinder functions as trial functions, Matsuhara (1973) has obtained propagation constants for dielectric waveguides with rectangular cross section shape and several inhomogeneous dielectric constant distributions.

Geshiro *et al.* (1974) have used vector variational expressions to study the effect of the ∇K term on the propagation constants of the wave modes. They have also considered a dielectric slab waveguide with truncated parabolic refractive index distribution.

Dispersion in Optical Waveguides

7.1. Introduction

By far the most important characteristic of a fiber, for its use in optical communications, is the spread of a pulse as it propagates through the fiber. The smaller the pulse dispersion, the greater the information carrying capacity. For example, if the pulse spread is reduced from 1000 to 1 nsec, the transmission capacity would increase from 1 million bit/sec to 1 billion bit/sec. Consequently, a careful analysis of the dispersion of a pulse in an optical fiber is of great importance in its potential use in optical communication.

   In order to understand the physical reasoning behind pulse dispersion we consider a ray which is launched into a cladded multimode fiber (of the type discussed in Chapter 4) making an angle θ with the axis; this angle θ is such that the ray periodically undergoes total internal reflections at the core–cladding interface (see Fig. 7.1). For total internal reflections to occur, θ should be less than $\pi/2 - \sin^{-1}(n_2/n_1)$, where n_1 and n_2 are the refractive indices of core and cladding, respectively. Obviously if L is the length of the fiber, the actual distance traversed by the ray is $L/\cos \theta$. Consequently, the propagation time of the ray in a fiber of length L will depend on the angle θ and would be given by

$$t(\theta) = \frac{n_1 L}{c \cos \theta} \qquad \text{for } \theta \leqslant \tfrac{1}{2}\pi - \sin^{-1}(n_2/n_1) \qquad (7.1)$$

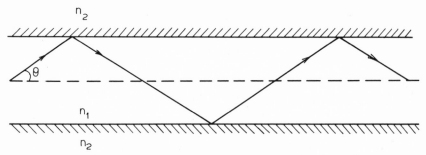

Fig. 7.1. Propagation of a ray in a cladded multimode fiber $(n_1 > n_2)$.

Thus the time taken by the rays to reach the end of the fiber increases with increasing θ which gives rise to dispersion. Since the minimum and maximum values of θ are 0 and $(\pi/2) - \sin^{-1}(n_2/n_1)$, the time difference between the longest and the shortest path lengths would be

$$
\begin{aligned}
\Delta &= \frac{n_1 L}{c \, \cos[(\pi/2) - \sin^{-1}(n_2/n_1)]} - \frac{n_1 L}{c} \\
&= \frac{n_1}{cn_2}(n_1 - n_2)L
\end{aligned}
\tag{7.2}
$$

The smaller the difference between the refractive indices of the core and cladding the smaller the time delay.* For $n_1 = 1.60$ with an index difference of 1% one obtains a time spread of about 50 nsec/km. In this case the bandwidth of a 1-km fiber is about 20 MHz.

Even under single mode operation† the bandwidth is finite, owing to the dispersion of the normal mode itself. Since the light frequencies are very large, the light pulses will invariably be long compared to a light cycle ($\sim 10^{-14}$ sec). Such pulses are not affected when $\beta - \omega$ characteristics are linear. However, the second derivative $d^2\beta/d\omega^2$ is important and is the cause of pulse distortions. Physically, it is the variation of group velocity within the modulation band that distorts the signal envelope. It can be shown that a Gaussian temporal pulse of initial width t_i spreads to [see Eq. (7.43)]

$$
t_0 = \left[t_i^2 + \left(\frac{L}{t_i} \frac{d^2\beta}{d\omega^2} \right)^2 \right]^{1/2}
\tag{7.3}
$$

---

*In the language of wave optics, the smaller the value of $n_1 - n_2$ the less the number of propagating modes through the fiber (see Fig. 4.4).

†The single mode condition for a simple cladded fiber can be written in the form [see Eq. (4.59)] $k_0 a(n_1^2 - n_2^2)^{1/2} < 2.4048$ where $k_0 = 2\pi/\lambda_0$, λ_0 being the free space wavelength, and a being the radius of the fiber. Thus, if the values of various parameters are such that the above inequality is satisfied, only one propagating mode will be excited.

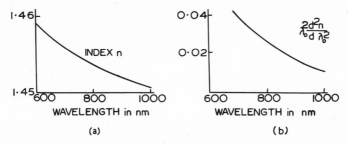

Fig. 7.2. Variation of the refractive index n and $\lambda_0^2(d^2n/d\lambda_0^2)$ with wavelength for silica (after Gloge, 1974; reprinted by permission).

in a dispersive medium of length L (see, for example, Garrett and McCumber, 1970; Gloge, 1970). The double derivative $d^2\beta/d\omega^2$ can be computed from Fig. 4.7. It should be noted that in the figure, β/k_0 is plotted as a function of $k_0a(n_1^2-n_2^2)^{1/2}$, the latter being proportional to the frequency. Numerical calculations for a typical fiber under single mode operation show that for $L = 1$ km an 8-psec pulse widens to 12 psec (Gloge, 1970). The corresponding bandwidth would be about 50 GHz.

Spreading of individual pulses can also occur on account of the wavelength dependence of the refractive index. Figure 7.2a shows the dependence of the refractive index of silica on the wavelength. The pulse spread (per unit distance) due to the dependence of n on wavelength is given by [see Eq. (7.69)]

$$\text{Pulse spread} = \frac{B}{c}\lambda_0^2\frac{d^2n}{d\lambda_0^2} \qquad (7.4)$$

where B represents the relative bandwidth ($= \Delta\lambda/\lambda$) of the source. For example, for a 1-nsec pulse produced by an LED, the carrier components are spread over a wavelength range of 360 Å and the corresponding pulse spread is about 3.6 nsec/km (Gloge, 1974), which agrees with the experimental data of Gloge *et al.* (1974). Similarly, for the GaAs laser, assuming a 20-Å bandwidth, one obtains a spread of about 0.2 nsec/km.

7.2. Impulse Response for a Gaussian Source in a Cladded Fiber–Ray Treatment*

Let a delta function impulse of energy E_0 be launched at one end of a cladded fiber at $t = 0$. If $E(\theta)\,d\theta$ represents the energy contained in the

*Reference: Dakin *et al.* (1973).

hollow annular cone lying between the angles θ and $\theta + d\theta$, then

$$E_0 = \int_0^{\pi/2} E(\theta)\, d\theta \tag{7.5}$$

Since the time taken by a particular ray to reach the other end of the fiber depends on the angle θ [see Eq. (7.1)], the energy injected in the annular conical region bounded by the angles θ and $\theta + d\theta$ will arrive at the other end of the fiber in the time interval $dt = t(\theta + d\theta) - t(\theta)$. We are assuming no mode conversion so that each value of θ corresponds to a unique value of t. Thus, if $T(\theta)$ is the ray transmission factor then the power P_0 arriving at the output end of the fiber during the time interval $t(\theta + d\theta) - t(\theta)$ is given by

$$
\begin{aligned}
P_0 &= \frac{E(\theta)T(\theta)\, d\theta}{t(\theta + d\theta) - t(\theta)} \\
&= \frac{E(\theta)T(\theta)}{dt/d\theta} = \frac{2\pi I(\theta)T(\theta)\sin\theta}{dt/d\theta}
\end{aligned}
\tag{7.6}
$$

where we have introduced the angular flux density, $I(\theta)$, which is related to $E(\theta)$ by the following

$$E(\theta)\, d\theta = 2\pi I(\theta)\sin\theta\, d\theta \tag{7.7}$$

where $2\pi\sin\theta\, d\theta$ is the solid angle between the two cones of semiangle θ and $\theta + d\theta$. It might be worthwhile to mention that the units of E_0, $E(\theta)$, $I(\theta)$, and $P_0(\theta)$ are joules, joules per radian, joules per steridian, and watts, respectively. Substituting the expression for $dt/d\theta$ from Eq. (7.1), we obtain

$$
P_0(t) = \frac{2\pi c}{n_1 L} I[\theta(t)] T[\theta(t)] \cos^2\theta(t) \qquad \text{for } \frac{n_1^2 L}{cn_2} \geq t \geq \frac{n_1 L}{c}
$$

$$
= 0 \qquad \text{for } t < \frac{n_1 L}{c} \text{ and } t > \frac{n_1^2 L}{cn_2}
\tag{7.8}
$$

where

$$\theta(t) = \cos^{-1}\left(\frac{n_1 L}{ct}\right) \tag{7.9}$$

We will apply the above results for an incident beam having a Gaussian intensity distribution. We have shown in Chapter 5 that in a uniform homogeneous medium, a Gaussian beam remains Gaussian, with its beam waist increasing with z [see Eqs. (5.90) and (5.104)]. Thus the intensity distribution could be written as

$$\frac{I_0}{\pi w^2(z)} \exp\left[-\frac{r^2}{w^2(z)}\right] \tag{7.10}$$

where

$$w^2(z) = w_0^2\left(1 + \frac{z^2}{k^2 w_0^4}\right)$$

Now if $I(r)\,dr$ represents the energy passing through an annular ring between r and $r + dr$ then

$$I(r)\,dr = \frac{I_0}{\pi w^2} 2\pi r \exp\left(-\frac{r^2}{w^2}\right) dr = E(\theta)\,d\theta \tag{7.11}$$

Notice that

$$\int_0^\infty I(r)\,dr = I_0 \qquad \text{(independent of } z) \tag{7.12}$$

For $z \gg k w_0^2$, we obtain approximately

$$w(z) \approx z/k w_0 \tag{7.13}$$

showing that, in the far-field pattern, the beam waist increases linearly with z. Now, curve 1 in Fig. 7.3 shows the locus of the point where the intensity falls to e^{-2} of the value at $r = 0$ [this will occur at $r = \sqrt{2}w(z) \approx (\sqrt{2}z/k w_0)$]. Thus, if the locus (for large z) makes an angle θ_0 with the z axis then

$$\tan\theta_0 = \sqrt{2}/k w_0 \tag{7.14}$$

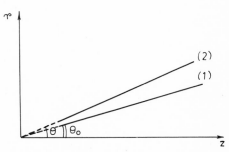

Fig. 7.3. Curve 1 is the locus of the points where the intensity falls to $1/e^2$ of the value at $r = 0$ for an incident beam having Gaussian spatial variation. Notice that in the far-field pattern the width increases linearly with z. Curve 2 represents a general ray.

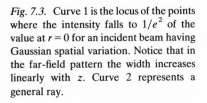

In general

$$r = z \tan \theta \tag{7.15}$$

so that

$$\frac{dr}{d\theta} = z \sec^2 \theta \tag{7.16}$$

Substituting this in Eq. (7.11) we obtain

$$E(\theta) \approx \frac{2I_0}{\pi(z^2/k^2 w_0^2)} z \tan \theta \exp\left(-\frac{r^2 k^2 w_0^2}{z^2}\right) z \sec^2 \theta$$

$$= \frac{4I_0 \tan \theta}{\tan^2 \theta_0 \cos^2 \theta} \exp\left(-\frac{2 \tan^2 \theta}{\tan^2 \theta_0}\right) \tag{7.17}$$

Substituting the above expression in Eq. (7.8) we obtain

$$P_0(t) = \left(\frac{2c}{n_1 L \tan \theta_0}\right)^2 t \exp\left\{-\frac{2}{\tan^2 \theta_0}\left[\left(\frac{ct}{n_1 L}\right)^2 - 1\right]\right\} T[\theta(t)]$$

$$\text{for } \frac{n_1 L}{c} \leqslant t \leqslant \frac{n_1^2 L}{cn_2} \tag{7.18}$$

$$= 0 \qquad\qquad \text{for } t < \frac{n_1 L}{c} \text{ and } t > \frac{n_1^2 L}{cn_2}$$

Assuming a lossless fiber (i.e., $T[\theta(t)] = 1$) with no mode conversion, a general shape of the impulse response is shown in Fig. 7.4. The leading edge appears at $t = t_1 (= n_1 L/c)$; the corresponding power being given by

$$P_0(t_1) = \frac{4c}{n_1 L \tan^2 \theta_0} \tag{7.19}$$

The pulse ends at $t = t_2 (= n_1^2 L/n_2 c)$ and at $t = t_m (= n_1 L \tan \theta_0/2c)$ the power attains its maximum value given by

$$P_0(t_m) = \frac{2c}{n_1 L \tan \theta_0} \exp\left(\frac{2}{\tan^2 \theta_0} - \frac{1}{2}\right) \tag{7.20}$$

However, the parameters may be such that t_1 is greater than t_m and a maximum is not observed.

Dakin *et al.* (1973) have also derived the impulse response for a lambertian source characterized by

$$I(\theta) = (1/\pi) \cos \theta \tag{7.21}$$

Such distributions are indeed produced by a light-emitting diode or other nonlasing sources. For such a source, the impulse response is given by

$$P_0(t) = 2\frac{(n_1 L/c)^2}{t^3} \qquad \text{for } \frac{n_1 L}{c} \leqslant t \leqslant \frac{n_1^2 L}{cn_2}$$

$$= 0 \qquad \text{for } t < \frac{n_1 L}{c} \quad \text{and} \quad t > \frac{n_1^2 L}{cn_2} \tag{7.22}$$

7.3. Gaussian Temporal and Spatial Distribution

Once the impulse response is known, the output pulse shape $G_0(t)$ can be determined for an arbitrary input pulse shape $G_i(t)$, by using the following relation:

$$G_0(t) = \int_{\tau_1}^{\tau_2} P_0(t-\tau)G_i(\tau)\, d\tau \tag{7.23}$$

where

$$\tau_1 = t - \frac{n_1^2 L}{cn_2}$$

$$\tau_2 = t - \frac{n_1 L}{c} \tag{7.24}$$

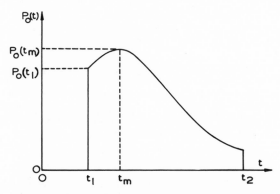

Fig. 7.4. Impulse response for a source with a Gaussian spatial distribution as given by Eq. (7.18) (after Dakin *et al.*, 1973; reprinted by permission).

The pulses produced by a mode-locked laser have approximately Gaussian temporal as well as spatial dependences; thus for such pulses we may write

$$G_i(t) = G_1 \exp\left(-\frac{t^2}{2t_i^2}\right) \qquad (7.25)$$

where

$$\int_{-\infty}^{+\infty} G_i(t)\, dt = E_0 \qquad (7.26)$$

Using Eqs. (7.18), (7.23), and (7.25) the output pulse shape is given by (Dakin *et al.*, 1973)

$$G_0(t) = 2\alpha G_1 \exp\left[\alpha\left(1 - \frac{t^2}{t_i^2\beta^2\gamma}\right)\right]$$

$$\times \int_1^{n_1/n_2} x \exp\left[-\gamma\left(\beta x - \frac{t}{t_i^2\gamma}\right)^2\right] T\left[\theta\left(\frac{n_1 Lx}{c}\right)\right] dx \qquad (7.27)$$

where

$$\alpha = \frac{2}{\tan^2 \theta_0}$$

$$\beta = \frac{n_1 L}{c}$$

$$\gamma = \frac{\alpha}{\beta^2} + \frac{1}{t_i^2}$$

Numerical calculations for typical values of various parameters have been carried out by Gambling *et al.* (1972).

7.4. Ray Delay in Graded Index Fibers

For a SELFOC fiber characterized by the refractive index variation

$$n(r) = n_0(1 - \tfrac{1}{2}\alpha^2 r^2 + \tfrac{1}{2}\beta'\alpha^4 r^4 + \cdots) \qquad (7.28)$$

Bouillie *et al.* (1974) have shown that, for meridional rays,* the time taken by a ray to traverse a fiber of length L is given by

$$t(\theta) = t_0[1 + \tfrac{3}{16}(\tfrac{5}{12} - \beta')\theta^4] \qquad (7.29)$$

where θ is the angle that the ray makes with the axis of the fiber (see Fig. 7.5) and $t_0 = (L/c)n_0$ represents the delay of the axial ray. Notice the weak

---

*Meridional rays are always confined to a single plane which contains the z axis.

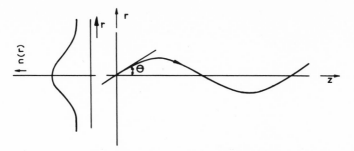

Fig. 7.5. Ray path in the graded index media.

dependence of t on θ. Further, for $\beta' = \frac{5}{12}$, $t(\theta)$ is independent of θ and therefore the refractive index distribution is known as a meridionally exact distribution [see Eq. (9.99)]. For the helically exact distribution [see Eq. (9.100)], $\beta' = \frac{3}{8}$ and

$$\Delta t = t(\theta) - t_0 = -\frac{1}{16}t_0\theta^4 \tag{7.30}$$

Once the θ dependence of t is known one can calculate the impulse response in a manner similar to the one described above (see Bouillie *et al.*, 1974).

7.5. *Broadening of a Gaussian Temporal Pulse using Modal Analysis**

In this section we will consider the broadening of a Gaussian temporal pulse without resorting to the geometrical optics approximation and accounting for the effects of material dispersion.

If the transverse spatial variation of the pth mode in a waveguide is denoted by $E_p(x, y)$, then the field pattern as it propagates along the z axis of the fiber is given by

$$E_p(x, y, z) = E_p(x, y) \exp(-i\beta_p z) \tag{7.31}$$

where the time dependence is assumed to be of the form $\exp(i\omega t)$. The above equation may be rewritten in the form

$$E_p(x, y, z) = E_p(x, y, 0)S_p(\omega)$$

where

$$S_p(\omega) = \exp(-i\beta_p z) \tag{7.32}$$

*Reference: Kapron and Keck (1971) and Gambling and Matsumura (1973).

represents the amplitude transfer function (ATF) at the frequency ω. We assume that an incident pulse, whose time dependence is described by the function $f(t)$ is incident on the entrance aperture of the guide (i.e., $z = 0$). Each component of its Fourier transform

$$F(\omega) = \frac{1}{\sqrt{2\pi}} \int_{-\infty}^{+\infty} f(t) \exp(-i\omega t)\, dt \qquad (7.33)$$

is assumed to be independently acted upon by the ATF to give the following expression for the output pulse

$$q_p(t) = \frac{1}{\sqrt{2\pi}} \int_{-\infty}^{+\infty} F(\omega) S_p(\omega) \exp(i\omega t)\, d\omega \qquad (7.34)$$

For a Gaussian temporal pulse, $f(t)$ is given by

$$f(t) = \exp\left(-\frac{t^2}{2t_i^2} + i\omega_c t \right) \qquad (7.35)$$

where ω_c represents the carrier frequency, and t_i is a measure of the width of the input pulse [see Eq. (7.58)]. As already mentioned, the temporal distribution of mode-locked pulses from a helium–neon laser is indeed Gaussian. The Fourier transform of $f(t)$ can easily be determined and is given by

$$F(\omega) = t_i \exp\left[-\frac{(\omega - \omega_c)^2 t_i^2}{2} \right] \qquad (7.36)$$

If we substitute the above expression for $F(\omega)$ in Eq. (7.34) and use Eq. (7.32) we would obtain

$$q_p(t) = \frac{1}{\sqrt{2\pi}} t_i \int_{-\infty}^{+\infty} \exp[-\tfrac{1}{2}(\omega - \omega_c)^2 t_i^2] \exp[i(\omega t - \beta_p z)]\, d\omega \qquad (7.37)$$

The function $F(\omega)$ is, in general, a very sharply peaked function around $\omega = \omega_c$. Even for a 10-psec pulse (i.e., $t_0 \approx 10$ psec), $F(\omega)$ would be negligible for $|\omega - \omega_c| \gtrsim 5/t_0 \approx 5 \times 10^{11}\ \text{sec}^{-1}$ (one must remember that $\omega_c \approx 3 \times 10^{15}\ \text{sec}^{-1}$). For larger values of t_0, the function $F(\omega)$ will be more sharply peaked. Consequently, we will be justified in expanding β_p around $\omega = \omega_c$ and retaining terms only up to the second order:

$$\beta_p(\omega) = \sum_{l=0}^{2} \frac{(\omega - \omega_c)^l}{l!} \alpha_{l,p}$$
$$= \alpha_{0,p} + (\omega - \omega_c)\alpha_{1,p} + \frac{(\omega - \omega_c)^2}{2!} \alpha_{2,p} \qquad (7.38)$$

where

$$\alpha_{l,p} = \frac{d^l}{d\omega^l}\beta_p(\omega)\Bigg|_{\omega=\omega_c} \tag{7.39}$$

Thus

$$q_p(t) = \frac{t_i}{\sqrt{2\pi}}\exp[i(\omega_c t - \alpha_{0,p}z)]\int \exp\left[-\left(\frac{t_i^2}{2} + \frac{i\alpha_{2,p}z}{2}\right)\Omega^2 - (i\alpha_{1,p}z)\Omega\right]d\Omega$$

$$= \frac{t_i \exp[i(\omega_c t - \alpha_{0,p}z)]}{(t_i^2 + i\alpha_{2,p}z)^{1/2}}\exp\left[-\frac{(t - z\alpha_{1,p})^2}{2(t_i^2 + iz\alpha_{2,p})}\right] \tag{7.40}$$

where $\Omega = \omega - \omega_c$. The shape of the detected output pulse, contributed by the pth mode, would be given by

$$|q_p(t)|^2 = \left[1 + \left(\frac{z\alpha_{2,p}}{t_i^2}\right)^2\right]^{-1/2}\exp\left\{-\frac{(t - z\alpha_{1,p})^2}{t_i^2\left[1 + \left(\frac{z\alpha_{2,p}}{t_i^2}\right)^2\right]}\right\} \tag{7.41}$$

From the above equation we make the following observations.

1. The maximum of the output pulse occurs at

$$t = z\alpha_{1,p} = z\frac{d\beta_p}{d\omega} \tag{7.42}$$

This is also to be physically expected because $d\beta_p/d\omega$ represents the group velocity of the mode and $z\, d\beta_p/d\omega$ would represent the time taken by the mode to traverse a distance z. Further, if higher order derivatives (like $\alpha_{2,p}$) are neglected, Eq. (7.41) predicts an undistorted pulse except for the time delay.

2. Second and higher order derivatives are responsible for the distortion of the pulse. Indeed if third and higher order derivatives are neglected then we find that a Gaussian pulse remains Gaussian with its width increasing from t_i to t_o where

$$t_o^2 = t_i^2\left[1 + \left(\frac{z\alpha_{2,p}}{t_i^2}\right)^2\right]$$

or

$$t_o = \left[t_i^2 + \left(\frac{z}{t_i}\frac{d^2\beta_p}{d\omega^2}\right)^2\right]^{1/2} \tag{7.43}$$

where the subscript o refers to the output pulse.

7.6. Gaussian Temporal and Spatial Pulse in a SELFOC Fiber

Gambling and Matsumura (1973) have studied the broadening of a Gaussian temporal pulse having a Gaussian spatial distribution in a SELFOC fiber. They have assumed the dielectric constant variation of the form

$$K(r) = K_0[1 - g^2 r^2 + b(gr)^4] \qquad (7.44)$$

where b and g are assumed to be constants. Thus the scalar wave equation, for azimuthally symmetric fields, is given by

$$\frac{1}{r}\frac{d}{dr}\left(r\frac{dE}{dr}\right) + \{k^2[1 - (gr)^2 + b(gr)^4] - \beta^2\}E = 0 \qquad (7.45)$$

where $k^2 \equiv (\omega^2/c^2)K_0$ and the z and time dependence is assumed to be of the form $\exp[i(\omega t - \beta z)]$. For $b = 0$, the modes can easily be determined* and are given by

$$E_p(r) = \frac{1}{\sqrt{\pi}w_0}L_p\left(\frac{r^2}{w_0^2}\right)\exp\left(-\frac{r^2}{2w_0^2}\right) \qquad (7.46)$$

where $L_p(x)$ is a Laguerre polynomial of degree p and $w_0^2 = 1/gk$. For example,

$$L_0(x) = 1; \quad L_1(x) = 1 - x; \quad L_2(x) = 1 - 2x + \tfrac{1}{2}x^2, \quad \text{etc.} \qquad (7.47)$$

The corresponding propagation constants are given by†

$$\beta_p^{(0)} = [k^2 - 2(2p+1)gk]^{1/2} \qquad p = 0, 1, 2, \ldots \qquad (7.48)$$

where the superscript (0) reflects the fact that we have neglected the quartic term in the dielectric constant variation. If we take into account this term, then in the first order perturbation theory (see Sec. 6.5) the propagation constants modify to (Gambling and Matsumura, 1973)

$$\beta_p = k\left[1 - \frac{2(2p+1)g}{k} + 2(3p^2 + 3p + 1)b\left(\frac{g}{k}\right)^2\right]^{1/2} \qquad (7.49)$$

---

*The procedure is similar to the one followed in chapter 5. One assumes a solution of the form $E = L(r)\exp(-r^2/2w_0^2)$ and the differential equation for $L(r)$ is solved by the series solution method. The infinite series behaves as $\exp(r^2/w_0^2)$ for large values of r and therefore the fields would diverge for $r \to \infty$ unless the series is terminated. The latter procedure leads to discrete values for β with corresponding specific field patterns.

†It should be pointed out that the complete modal analysis (using the scalar wave equation) for the parabolic dielectric constant variation has been carried out in Chapter 5. Here we have just picked out those modes which have azimuthal symmetry (i.e., independent of φ coordinate). The fundamental mode ($p = 0$) corresponds to $n = m = 0$ [see Eq. (5.18); K_2 must be set equal to K_3]. The $p = 1$ mode is a linear combination of the $n = 1$, $m = 0$ and $n = 0$, $m = 1$ modes, etc.

Next, we assume that the input beam has Gaussian spatial distribution, i.e.,

$$E(r, 0) = \frac{1}{\sqrt{\pi} w_i} \exp\left[-\frac{r^2}{2w_i^2}\right] \tag{7.50}$$

If the beam is launched centrally along the axis of the fiber, then only the azimuthally symmetric modes will be excited and we may write

$$\frac{1}{\sqrt{\pi} w_i} \exp\left(-\frac{r^2}{2w_i^2}\right) = \sum_{p=0}^{\infty} A_p \left[\frac{1}{\sqrt{\pi} w_0} L_p\left(\frac{r^2}{w_0^2}\right) \exp\left(-\frac{r^2}{2w_0^2}\right)\right] \tag{7.51}$$

Using the orthonormal properties of the Laguerre–Gauss functions we obtain

$$\begin{aligned}
A_p &= \frac{2\sqrt{\pi}}{w_0} \int_0^{\infty} E(r, 0) L_p\left(\frac{r^2}{w_0^2}\right) \exp\left(-\frac{r^2}{2w_0^2}\right) r \, dr \\
&= \frac{2W}{1+W^2}\left(\frac{W^2-1}{W^2+1}\right)^p
\end{aligned} \tag{7.52}$$

where $W = w_0/w_i$. Note that if $W = 1$ (i.e., $w_i = w_0$), $A_p = 0$ for $p \neq 0$ which implies that only the fundamental mode is excited. In general, for an arbitrary intensity distribution of the incident beam, all the modes will be excited and each mode will propagate independently through the fiber—we have, of course, assumed no mode conversion (see Chapter 8). For a temporal pulse, the time variation of each mode will become distorted [see Eq. (7.41)]; and the temporal distribution of the output pulse would be given by

$$|q(t)|^2 = \sum_{p=0}^{\infty} |A_p|^2 |q_p(t)|^2 \int_0^{\infty} \frac{1}{\pi w_0^2}\left[L_p\left(\frac{r^2}{w_0^2}\right) \exp\left(-\frac{r^2}{2w_0^2}\right)\right]^2 2\pi r \, dr \tag{7.53}$$

where we have assumed that the detector (placed at the output end of the fiber) receives the entire energy coming out of the fiber. Carrying out the integration we get

$$\begin{aligned}
|q(t)|^2 &= \sum_{p=0}^{\infty} |A_p|^2 |q_p(t)|^2 \\
&= \frac{4W^2}{(1+W^2)^2} \sum_{p=0}^{\infty} \left[\frac{W^2-1}{W^2+1}\right]^{2p} \left[1+\left(\frac{z\alpha_{2,p}}{t_i^2}\right)^2\right]^{-1/2} \\
&\qquad\qquad\qquad\qquad \times \exp\left\{-\frac{(t-z\alpha_{1,p})^2}{t_i^2\left[1+\left(\frac{z\alpha_{2,p}}{t_i^2}\right)^2\right]}\right\}
\end{aligned} \tag{7.54}$$

Further, using the expression for β [see Eq. (7.49)] we obtain

$$\alpha_{1,p} = \frac{d\beta_p}{d\omega}\bigg|_{\omega=\omega_c}$$

$$= \left\{ \left[\frac{dk}{d\omega} - (2p+1)\frac{dg}{d\omega} \right] + \left[2(1-\frac{3b}{2})(p^2+p) + (\tfrac{1}{2}-b) \right] \right.$$

$$\left. + \left[\frac{g}{k}\left(\frac{g}{k}\frac{dk}{d\omega} - 2\frac{dg}{d\omega} \right) \right] \right\}_{\omega=\omega_c} \qquad (7.55)$$

and

$$\alpha_{2,p} = \frac{d^2\beta_p}{d\omega^2}\bigg|_{\omega=\omega_c}$$

$$= \left\{ \left[\frac{d^2k}{d\omega^2} - (2p+1)\frac{d^2g}{d\omega^2} \right] + [2(1-\tfrac{3}{2}b)(p^2+p) + (\tfrac{1}{2}-b)] \right.$$

$$\left. + \left[\frac{g^2}{k^2}\frac{d^2k}{d\omega^2} - \frac{2g}{k}\frac{d^2g}{d\omega^2} + \frac{4g}{k^2}\frac{dk}{d\omega}\frac{dg}{d\omega} - \frac{2g^2}{k^3}\left(\frac{dk}{d\omega}\right)^2 - \frac{2}{k}\left(\frac{dg}{d\omega}\right)^2 \right] \right\}_{\omega=\omega_c}$$

$$(7.56)$$

Now, the dispersion of a waveguide is usually defined as the difference in the widths of the input and output pulse; and by the width of a pulse we imply its width at half maximum. Thus, for a Gaussian temporal distribution whose intensity varies as $\exp(-t^2/t_i^2)$, the width of the pulse would be $2\tau_i$ where

$$\exp\left(-\frac{\tau_i^2}{t_i^2} \right) = \frac{1}{2} \qquad (7.57)$$

Thus the width of the input pulse would be given by

$$2\tau_i = 2t_i(\ln 2)^{1/2} \qquad (7.58)$$

When the output pulse is also Gaussian, the dispersion, Δ, would be given by

$$\Delta = 2\tau_0 - 2\tau_i \qquad (7.59)$$

where $2\tau_0$ is the width of the output pulse. In particular if $W=1$, i.e., only the fundamental mode is excited and the dispersion would be given by

$$\Delta = 2\tau_0 - 2\tau_i$$

$$= 2t_i(\ln 2)^{1/2}\left\{ \left[1 + \left(\frac{z\alpha_{2,0}}{t_i^2} \right)^2 \right]^{1/2} - 1 \right\} \qquad (7.60)$$

which shows that for large values of z, the dispersion would linearly increase with z. For a single mode excitation (i.e., $W = 1$) Fig. 7.6 shows the variation of the pulse dispersion, Δ, with the parameter b which represents the strength of the quartic term. The values of various parameters are as follows: $g = 1.77$ mm^{-1} (independent of ω), $\omega_c = 2.975 \times 10^{15}$ sec^{-1}, width of the input pulse, $2\tau_i = 0.65$ nsec. The derivatives $dn_0/d\omega$ and $d^2n_0/d\omega^2$ are calculated from Fig. 7.7. Notice that in a SELFOC fiber the pulse broadening is predominantly due to material dispersion. When n_0 is assumed to be independent of ω the dispersion decreases by a factor of about 10^{-10} and is almost zero for $b \approx 0$. Physically, this corresponds to the fact that the modes have almost the same group velocity [see Eq. (5.20)].

For $W \neq 1$ one has to sum the entire series on the RHS of Eq. (7.54) and the corresponding variation of pulse dispersion with W for $b = 1$ is shown in Fig. 7.8. From the figure we notice that minimum dispersion occurs for $W = 1$.

Marcuse (1973e) has calculated the width of the impulse response function of a parabolic index fiber with finite radius. He assumes that all modes carry equal amounts of power up to a certain maximum mode

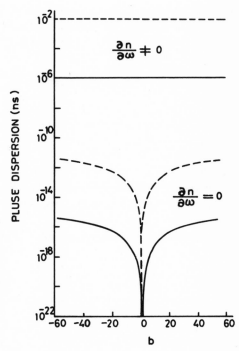

Fig. 7.6. Variation of pulse dispersion with parameter b. The values of other parameters are given by Eq. (7.61). The lower two curves correspond to $dn/d\omega = 0$ and the upper two curves take into account material dispersion as given by Fig. 7.7. The solid and dashed curves correspond to $z = 1$ and $z = 100$ km, respectively (after Gambling and Matsumura, 1973; reprinted by permission).

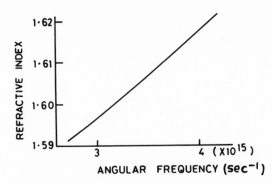

Fig. 7.7. The dependence of n on ω for the SELFOC glass (after Gambling and Matsumura, 1973; reprinted by permission).

number which is determined by the cutoff condition that all modes which satisfy the inequality

$$\beta > n(a)k_0 \qquad (7.62)$$

do not propagate through the fiber. Here a represents the radius of the core. Assuming no material dispersion, the impulse response comes out to be rectangular, the width of which is given by

$$d\tau = \frac{K_o^{1/2}L}{2c} \frac{ag}{\sqrt{2}} \qquad (7.63)$$

where L is the length of the fiber and g the grading parameter defined by Eq. (7.44).

However, an impulse (which is a delta function in time) is really an idealized concept. In fact, for an ideal impulse, all frequencies are equally excited and one can not talk of a carrier frequency. Consequently what one really should calculate is the broadening of a sharp Gaussian pulse rather than an impulse response. Even for a pulse of as short a duration as 10 psec, there will be about 10,000 oscillations and one will be well justified to make the expansion expressed by Eq. (7.38). Thus, when one is carrying out a wave optics treatment one should not calculate the impulse response (which will lead to unnecessary mathematical difficulties*); instead the response of a Gaussian pulse should be obtained as has indeed been done in this section.

---

*For a delta function in time, $F(\omega)$ will be independent of ω [see Eq. (7.33)] and one cannot make the expansion given by Eq. (7.38).

Returning to Eq. (7.56), we note that since

$$k = (\omega/c)K_0^{1/2} \tag{7.64}$$

we will have

$$\frac{dk}{d\omega} = \frac{n_0}{c} + \frac{\omega}{c}\frac{dn_0}{d\omega} \tag{7.65}$$

and

$$\frac{d^2k}{d\omega^2} = \frac{\omega}{c}\frac{d^2n_0}{d\omega^2} + \frac{2}{c}\frac{dn_0}{d\omega} \tag{7.66}$$

where $n_0 = K_0^{1/2}$ represents the refractive index on the axis. Now

$$\frac{dn_0}{d\omega} = \frac{dn_0}{d\lambda_0}\frac{d\lambda_0}{d\omega} = -\frac{2\pi c}{\omega^2}\frac{dn_0}{d\lambda_0} = -\frac{\lambda_0^2}{2\pi c}\frac{dn_0}{d\lambda_0}$$

and

$$\frac{d^2n_0}{d\omega^2} = -\left(\frac{\lambda_0}{\pi c}\frac{dn_0}{d\lambda_0} + \frac{\lambda_0^2}{2\pi c}\frac{d^2n_0}{d\lambda_0^2}\right)\left(-\frac{\lambda_0^2}{2\pi c}\right)$$

$$\therefore \frac{d^2k}{d\omega^2} = \frac{2\pi}{\lambda_0}\left(\frac{\lambda_0^2}{2\pi c}\right)\left[\frac{\lambda_0}{\pi c}\frac{dn_0}{d\lambda_0} + \frac{\lambda_0^2}{2\pi c}\frac{d^2n_0}{d\lambda_0^2}\right] - \frac{2}{c}\frac{\lambda_0^2}{2\pi c}\frac{dn_0}{d\lambda_0}$$

$$= \frac{\lambda_0}{2\pi c^2}\left(\lambda_0^2\frac{d^2n_0}{d\lambda_0^2}\right) \tag{7.67}$$

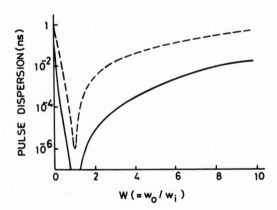

Fig. 7.8. The dependence of pulse dispersion on W for $b = 1$ and $z = 1$ km. The solid curve corresponds to $dn/d\omega = 0$ (after Gambling and Matsumura, 1973; reprinted by permission).

If we assume that the term $d^2k/d\omega^2$ (due to material dispersion) dominates in the expression for $\alpha_{2,p}$ [see Eq. (7.56)] then

$$\Delta \approx 2t_i (\ln 2)^{1/2} \left\{ \left[1 + \left(\frac{z\alpha_{2,0}}{t_i^2} \right)^2 \right]^{1/2} - 1 \right\}$$

$$\approx 2t_i (\ln 2)^{1/2} \frac{z}{t_i^2} \frac{\lambda_0}{2\pi c^2} \lambda_0^2 \frac{d^2 n_0}{d\lambda_0^2} \tag{7.68}$$

Now if the initial bandwidth is $\Delta\nu$ then

$$t_i \sim \frac{1}{\Delta\nu}$$

or, omitting the sign

$$t_i \sim \frac{\lambda_0^2}{c\Delta\lambda_0}$$

Thus

$$\Delta \approx \left[\frac{(\ln 2)^{1/2}}{\pi} \frac{B}{c} \left(\lambda_0^2 \frac{d^2 n_0}{d\lambda_0^2} \right) \right] z \tag{7.69}$$

where $B = \Delta\lambda_0/\lambda_0$ represents the relative bandwidth of the source. Equation (7.69) may be compared with Eq. (7.4).

7.7. Response of a Gaussian Pulse in a Cladded Fiber: Wave Optics Treatment*

In Secs. 7.2 and 7.3 we have calculated the response of a Gaussian temporal pulse in a cladded fiber using ray theory. In this section we will calculate the response using wave optics. In Chapter 4 we had mentioned that when v is large the values of u_m are given by [see Eq. (4.62)]

$$u_m = u_m(\infty) \exp(-1/v) \tag{7.70}$$

where $u_m(\infty)$ is the mth root of $J_l(u) = 0$. We assume that the incident pulse has Gaussian spatial distribution and is launched centrally along the axis of the fiber so that only the azimuthally symmetric modes are excited. Consequently, if we refer to Eq. (4.40) only the $l = 0$ modes will be excited whose

*Reference: Ghatak et al. (1975).

spatial distributions are given by

$$(E_y)_m = E_0 \frac{1}{J_0(u_m)} J_0\left(u_m \frac{r}{a}\right) \qquad \text{for } 0 < r < a$$

$$= E_0 \frac{1}{K_0(w_m)} K_0\left(w_m \frac{r}{a}\right) \qquad \text{for } r > a \tag{7.71}$$

where all symbols have been defined in Chapter 4. If we define

$$E_0 = \left\{ a(\pi n_1)^{1/2} \left(\frac{v}{u_m}\right) \left[\frac{K_1(w_m)}{K_0(w_m)}\right] \right\}^{-1} \tag{7.72}$$

then it can easily be shown that

$$n_1 \int_0^\infty (E_y)_m (E_y)_n \, 2\pi r \, dr = \delta_{mn} \tag{7.73}$$

The modes described by Eq. (7.71) are known as LP_{0m} modes*. For these modes $u_m(\infty)$ are 2.4048, 5.5201, 8.6537, ... for $m = 1, 2, 3, \ldots$, respectively. The corresponding propagation constants are given by [see Eq. (4.36) and Eq. (7.70)]

$$\beta_m^2 = k_0^2 n_1^2 - \frac{u_m^2(\infty)}{a^2} \exp\left(-\frac{2}{v}\right) \tag{7.74}$$

Or, since $k_0 = \omega/c$ and $v = k_0 a[(n_1^2 - n_2^2)]^{1/2} = (\omega/c) a (n_1^2 - n_2^2)^{1/2}$ we get

$$\alpha_{1,m} = \frac{d\beta_m}{d\omega} = \left\{ \frac{k_0}{c\beta_m} \left[n_1^2 - \frac{u_m^2(\infty)}{k_0^2 a^2} \frac{1}{v} \exp\left(-\frac{2}{v}\right) \right] \right\}_{\omega = \omega_c} \tag{7.75}$$

and

$$\alpha_{2,m} = \frac{d^2\beta_m}{d\omega^2} = \left(\frac{1}{\beta_m} \left\{ \frac{1}{c^2} \left[n_1^2 + \frac{2}{v}\left(1 - \frac{1}{v}\right) \frac{u_m^2(\infty)}{k_0^2 a^2} \exp\left(-\frac{2}{v}\right) \right] - \alpha_{1,m}^2 \right\} \right)_{\omega = \omega_c} \tag{7.76}$$

where we have neglected material dispersion.

In order to calculate the excitation efficiency we expand the input Gaussian beam, E_i, in the field patterns of the LP_{0m} modes

$$E_i = \left(\frac{2}{\pi n_1 \sigma^2}\right)^{1/2} \exp\left(-\frac{r^2}{\sigma^2}\right) = \sum_m A_m (E_y)_m \tag{7.77}$$

where the factor in front of the Gaussian is such that

$$n_1 \int_0^\infty E_i^2 \, 2\pi r \, dr = 1 \tag{7.78}$$

*According to the rigorous analysis described in Sec. 4.4 these modes are designated as HE_{1m} modes.

Using Eq. (7.73), the coefficient A_m can be easily calculated. The excitation efficiency, P_m, which is simply A_m^2 is therefore given by (Gambling $et\ al.$, 1972; Snyder, 1969);

$$P_m = A_m^2 = 2\left(\frac{u_m}{v}\frac{K_0(w_m)}{K_1(w_m)}\frac{2a}{\sigma}\left\{\int_0^1 \frac{J_0(u_m\rho)}{J_0(u_m)}\exp\left[-\frac{\rho^2}{(\sigma/a)^2}\right]\rho\,d\rho\right.$$
$$\left.+\int_1^\infty \frac{K_0(w_m\rho)}{K_0(w_m)}\exp\left[-\frac{\rho^2}{(\sigma/a)^2}\right]\rho\,d\rho\right\}\right)^2 \qquad (7.79)$$

where $\rho = r/a$, a being the core radius. The output power would be given by [Eq. (7.54)]:

$$P_{\text{output}} = \sum_m P_m \frac{1}{\left[1+\left(\frac{z\alpha_{2,p}}{t_i^2}\right)^2\right]^{1/2}}\exp\left\{-\frac{(t-z\alpha_{1,p})^2}{t_i^2[1+(z\alpha_{2,p}/t_i^2)^2]}\right\} \qquad (7.80)$$

We next apply the above theoretical results for practical fibers. We assume the values of various parameters to be as follows:

$$n_1 = 1.551$$
$$n_2 = 1.485$$
$$a = 28.5\ \mu \qquad (7.81)$$
$$k_0 = 10^7\ \text{m}^{-1}(\lambda_0 \sim 0.633\ \mu\text{m})$$

Length of the fiber, $z = 1$ km

$$\theta_0 = 5°$$

Thus

$$k_0 a = 285$$
$$v = k_0 a(n_1^2 - n_2^2)^{1/2} = 127.6 \qquad (7.82)$$

and

$$\sigma = (\lambda_0/\pi\theta_0 n_1) = 1.48\ \mu\text{m}$$

The above values correspond to the experiments reported by Gambling $et\ al.$ (1972). Assuming an initial pulse width* of $\Delta_i = 0.65$ nsec, the fractional increase in the square of the pulse width, $(z\alpha_{2,m}/t_i^2)^2$ [see Eq. (7.41)], of various modes is tabulated in Table 7.1. It can easily be seen that the

---

*Pulse width $= 2(\ln 2)^{1/2}t_i$; see Eq. (7.58).

Table 7.1. Relative Delay Time and Fractional Increase in Square of the Pulse Width of Various Modes for a Fiber Charac-terized by Eq. (7.80)

Mode number, m	Relative delay time, nsec $z\alpha_{1,m} - zn_1/c$	$(z\alpha_{2,m}/t_i^2)^2$	
		$\Delta = 0.65$ nsec	$\Delta = 10$ psec
1	0.074	1.0×10^{-13}	1.8×10^{-6}
2	0.391	2.8×10^{-12}	5.1×10^{-5}
3	0.960	1.7×10^{-11}	3.1×10^{-5}
4	1.783	5.9×10^{-11}	1.1×10^{-3}
5	2.860	1.5×10^{-10}	2.7×10^{-3}
6	4.192	3.3×10^{-10}	5.8×10^{-3}
7	5.778	6.2×10^{-10}	1.1×10^{-2}
8	7.620	1.1×10^{-9}	1.9×10^{-2}
9	9.718	1.8×10^{-9}	3.1×10^{-2}
10	12.074	2.7×10^{-9}	4.8×10^{-2}
11	14.689	4.0×10^{-9}	7.2×10^{-2}
12	17.562	5.8×10^{-9}	1.0×10^{-1}
13	20.697	8.0×10^{-9}	1.4×10^{-1}
14	24.094	1.1×10^{-8}	1.9×10^{-1}

distortion is negligible. Even for $\Delta_i = 10$ psec, the distortion is very small particularly for low order modes. The delay time of the various modes is also tabulated. One can easily see that at the output end of the fiber, the modes will not overlap and one would get a large number of pulses (each corresponding to a particular mode) as shown in the oscillatory curve of Fig. 7.9. The summation in Eq. (7.80) extends to $p = 20$, which carry more than 99% of the total power. On the other hand, using the ray theory results of Dakin et al. (1973), the shape of the output pulse is given by [see Eq. (7.27)]:

$$P_{output} = 2\alpha \exp\left[\alpha\left(1 - \frac{t^2}{t_i^2 \beta^2 \gamma}\right)\right] \int_1^{n_1/n_2} x \exp\left[-\gamma\left(\beta x - \frac{t}{t_i^2 \gamma}\right)^2\right] dx \qquad (7.83)$$

where the symbols have been defined in Sec. 7.3. The variation of the output power with time is shown as the dashed curve in Fig. 7.9. The values of various parameters are the same as those given in Eq. (7.81). (It may be noted that the ray theory results are independent of the values of k_0 and a.) Comparison of the two curves shows that the ray theory predictions are very much different from the more accurate wave optics results. Consequently, for the experimental configuration of Gambling et al. (1972), one must use the correct wave optics results. However, the reported agreement of the ray theory result with the experimental data is fortuitous and is possibly due to

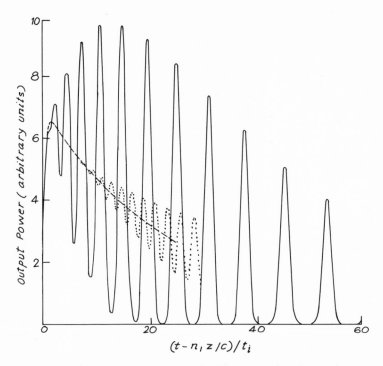

$(t - n_1 z /c)/t_i$

Fig. 7.9. The oscillatory curve corresponds to the dependence of the output power on time using the wave optics result [Eq. (7.90)]. The values of various parameters are the ones given in Eq. (7.81) with $\Delta = 0.65$ nsec. The corresponding ray theory results [as obtained from Eq. (7.83)] are shown as the dashed curve. The dashed curve also corresponds to the wave optics result [Eq. (7.80)] for $a = 200~\mu$ (i.e., $v = 895$ and $k_0 a = 2000$) with the same value of the other parameters. The dotted curve corresponds to $v = 300$ (after Ghatak *et al.*, 1975; reprinted by permission).

mode conversion; the latter also being responsible for a relatively smooth output.

It is also to be noted that if we take a very large value of v, the wave optics results match with the results obtained using ray theory. In the above example, the value of v is 127.6, which is not large enough for the ray theory to be valid. If, however, we take v to be 895 (which corresponds to $a = 200~\mu$ for $\lambda \simeq 0.63~\mu$), then the predictions of the wave optics theory match very well with that of the ray theory results. Indeed, the difference between the two results is so small that they cannot be shown on the scale of Fig. 7.9. The smooth variation of the output power (for very large values of v) is a manifestation of the fact that the delay difference between various modes decreases, and the pulses due to individual modes overlap. We have also made a calculation corresponding to an intermediate value of v ($= 300$), and the dotted curve in Fig. 7.9 corresponds to this value. The curve shows that, as v increases, the wave optics calculations approach the geometrical optics results. It may be mentioned that the effect of Goos Hanchen shift in the ray theory, is small (Gambling *et al.*, 1974; Sodha *et al.*, 1973). For an exponential profile (see Sec. 3.2), a similar analysis has been made by Khular and Ghatak (1976).

7.8. Dependence of Dispersion on Input Pulse Widths

Finally, we would like to emphasize that the dispersion (which is usually defined as the difference in the widths of the input and output pulses) is, in general, quite a sensitive function of the input pulse width; and one should be careful in calculating the information carrying capacity of a fiber. It often happens that for large input pulse widths the dispersion is extremely small; from this one should not infer a large information carrying capacity of the fiber. We will explicitly show this by considering two simple (but practical) examples. The treatment is based on a paper by Goyal *et al.* (1976).

For a given input pulse $I(t)$, the output pulse $O(t)$, would be given by

$$O(t) = \int P(t-\tau)I(\tau)\,d\tau \qquad (7.84)$$

where $P(t)$ is the impulse response function.

We first assume the impulse response function to be rectangular, i.e.,

$$\begin{aligned} P(t) &= P_0 && \text{for } t_1 < t < t_2 \\ &= 0 && \text{for } t < t_1 \text{ and } t > t_2 \end{aligned} \qquad (7.85)$$

Such a response is obtained for a parabolic index fiber of finite radius when all the propagating modes are equally excited (Marcuse, 1973e). Next, we assume the input pulse to be rectangular; i.e.,

$$\begin{aligned} I(t) &= I_0 && \text{for } 0 < t < t_0 \\ &= 0 && \text{for } t < 0 \text{ and } t > t_0 \end{aligned} \qquad (7.86)$$

where t_0 represents the width of the pulse. The output pulse shape is given by the following expressions:

(a) When $t_0 < (t_2 - t_1)$

$$\begin{aligned} O(t) &= P_0 I_0 (t - t_1) && \text{for } t_1 \leqslant t \leqslant t_1 + t_0 \\ &= P_0 I_0 t_0 && \text{for } t_1 + t_0 < t \leqslant t_2 \\ &= P_0 I_0 (t_2 + t_0 - t) && \text{for } t_2 < t \leqslant t_2 + t_0 \end{aligned} \qquad (7.87)$$

the width of which is simply $(t_2 - t_1)$ which is independent of t_0.

(b) When $t_0 > (t_2 - t_1)$

$$
\begin{aligned}
O(t) &= P_0 I_0(t - t_1) && \text{for } t_1 \leq t \leq t_2 \\
&= P_0 I_0(t - t_1) && \text{for } t_2 < t < t_1 + t_0 \qquad (7.88) \\
&= P_0 I_0(t_2 + t_0 - t) && \text{for } t_1 + t_0 < t < t_2 + t_0
\end{aligned}
$$

the width of which is equal to the input pulse width; thus the dispersion is zero.

We next assume the impulse response function to be Gaussian, i.e.,

$$
P(t) = P_0 \exp\left[-\frac{(t - \Delta)^2}{T^2} \right] \qquad (7.89)
$$

Such a response is expected when mode conversion occurs due to random irregularities in the fiber. For an input Gaussian pulse

$$
I(t) = I_0 \exp(-t^2/t_i^2) \qquad (7.90)
$$

one readily obtains

$$
O(t) = \frac{P_0 I_0 T \sqrt{\pi}}{[1 + (T^2/t_i^2)]^{1/2}} \exp\left[-\frac{(t - \Delta)^2}{T^2 + t_i^2} \right] \qquad (7.91)
$$

which is again a Gaussian distribution. The above equation tells us that the difference in the squares of output and input pulse widths is independent of the input pulse width! Using Eqs. (7.87), (7.88), or (7.91), one can easily calculate the dispersion as a function of input pulse width in the respective cases. Numerical calculations show that the dispersion is in general, quite sensitive to the input pulse width (see Goyal *et al.*, 1976).

CHAPTER 8

Propagation in Imperfect Waveguides

In a perfect waveguide the power associated with a particular mode remains constant as the beam propagates through the waveguide; we are, of course, neglecting the losses due to absorption. However, in general, there are imperfections in the waveguide; an important example of this is the irregular interface between the core and cladding or a weak z dependence of the dielectric constant. Such imperfections lead to transfer of power between the various modes of the waveguide and the power transferred to the radiation modes results in an exponential decay of the power.

In the first part of this chapter we will discuss the propagation in a waveguide whose dielectric constant variation can be written as a sum of two parts: one part depending only on the z coordinate and the other part depending on the transverse coordinates. It turns out that, under such conditions, it is not difficult to solve the scalar wave equation and one can obtain an analytical solution for some specific profiles. Although the treatment given corresponds to SELFOC fibers, it can be extended to other dielectric constant profiles.

In the second part we study the mode conversion in a SELFOC fiber $(K = K_0 - K_2 r^2)$ due to the z dependence of K_2. The treatment assumes an azimuthally symmetric beam incident on the entrance aperture of the fiber.

In the third part we develop the Green's function method to study the mode conversion in a waveguide whose dielectric constant variation can be

177

written as a sum of two parts: the first part corresponding to the perfect waveguide and the second part representing the variations due to various irregularities. We will follow the treatment of Ghatak *et al.* (1976), which is quite general and assumes that the effect due to the second term is small in comparison to the first term. The final result has been applied to an imperfect slab waveguide.

8.1. $K = K_0(z) + K_1(x, y)$

In this section we consider a dielectric constant variation of the form

$$K = K_0(z) + K_1(x, y) \tag{8.1}$$

where the propagation is along the z axis. Although the method described below is valid whenever the dielectric constant variation can be written as a sum of two parts, one depending on the z cordinate and the other depending on the transverse coordinates, we will, for preciseness, assume

$$K_1(x, y) = -K_2 x^2 - K_3 y^2 \tag{8.2}$$

where K_2 and K_3 are constants. Thus the scalar wave equation assumes the form

$$\nabla^2 \psi + (\omega^2/c^2)[K_0(z) - K_2 x^2 - K_3 y^2]\psi = 0 \tag{8.3}$$

Writing

$$\psi = X(x) Y(y) Z(z) \tag{8.4}$$

we readily obtain

$$\left(\frac{1}{X} \frac{d^2 X}{dx^2} - \frac{\omega^2}{c^2} K_2 x^2 \right) + \left(\frac{1}{Y} \frac{d^2 Y}{dy^2} - \frac{\omega^2}{c^2} K_3 y^2 \right) + \left(\frac{1}{Z} \frac{d^2 Z}{dz^2} + K_0(z) \frac{\omega^2}{c^2} \right) = 0 \tag{8.5}$$

For the fields to be bounded for $x, y \to \pm\infty$ we must have (see Chapter 5)

$$\frac{1}{X} \frac{d^2 X}{dx^2} - \frac{\omega^2}{c^2} K_2 \omega^2 = -(2n + 1) \frac{\omega}{c} K_2^{1/2} \tag{8.6a}$$

and

$$\frac{1}{Y} \frac{d^2 Y}{dy^2} - \frac{\omega^2}{c^2} K_3 y^2 = -(2m + 1) \frac{\omega}{c} K_3^{1/2} \tag{8.6b}$$

where $n, m = 0, 1, 2, \ldots$. Thus $Z(z)$ will satisfy

$$\frac{d^2Z}{dz^2} + \left[K_0(z)\frac{\omega^2}{c^2} - (2n+1)\frac{\omega}{c}K_2^{1/2} - (2m+1)\frac{\omega}{c}K_3^{1/2} \right]Z(z) = 0 \quad (8.7)$$

Equation (8.7) can easily be solved numerically for an arbitrary z dependence of $K_0(z)$. Analytic solutions are also possible for a variety of profiles. For example, if we choose

$$K_0(z) = K_{00} + \frac{b^2}{(z+z_o)^2} \quad (8.8)$$

we obtain

$$\frac{d^2Z}{d\zeta^2} + \left[1 + \frac{\omega^2}{c^2}\frac{b^2}{\zeta^2} \right]Z(\zeta) = 0 \quad (8.9)$$

where

$$\zeta = (z+z_0)\Lambda$$

$$\Lambda = \left[K_{00}\frac{\omega^2}{c^2} - (2n+1)\frac{\omega}{c}K_2^{1/2} - (2m+1)\frac{\omega}{c}K_3^{1/2} \right]^{1/2}$$

The solution of Eq. (8.9) is given by

$$Z(\zeta) = \zeta^{1/2}[AJ_p(\zeta) + BY_p(\zeta)] \quad (8.10)$$

where $J_p(\zeta)$ and $Y_p(\zeta)$ represent the Bessel and Weber functions and $p = [\frac{1}{4} - (\omega^2/c^2)b^2]^{1/2}$. The coefficients A and B can be related to each other by noting that for large values of z, $K_0(z)$ attains a constant value and hence for $z \to \infty$, we must have*

$$Z(z) \xrightarrow[z\to\infty]{} \exp(-i\beta z) \quad (8.11)$$

Now, using the asymptotic forms of $J_p(\zeta)$ and $Y_p(\zeta)$ we obtain

$$Z(\zeta) \xrightarrow[\zeta\to\infty]{} \zeta^{1/2}\left[A\left(\frac{2}{\pi\zeta}\right)^{1/2}\cos\left(\zeta - \frac{\pi}{4} - \frac{p\pi}{2}\right) + B\left(\frac{2}{\pi\zeta}\right)^{1/2}\sin\left(\zeta - \frac{\pi}{4} - \frac{p\pi}{2}\right) \right]$$

$$= \left(\frac{1}{2\pi}\right)^{1/2}\left\{ (A+B)\exp\left[i\left(\zeta - \frac{\pi}{4} - \frac{p\pi}{2}\right) \right] \right.$$

$$\left. + \left(A - \frac{b}{i}\right)\exp\left[-i\left(\zeta - \frac{\pi}{4} - \frac{p\pi}{2}\right) \right] \right\}$$

Using Eq. (8.11), we obtain

$$B = -iA$$

*We are assuming the time dependence to be of the form $\exp(+i\omega t)$.

Thus

$$Z(\zeta) = A\zeta^{1/2}[J_p(\zeta) - iY_p(\zeta)] \tag{8.12}$$

It is interesting to note that in the absence of any inhomogeneity in the z direction, $p = \frac{1}{2}$ and

$$Z(\zeta) = A\zeta^{1/2}\left[\left(\frac{2}{\pi\zeta}\right)^{1/2}\sin\zeta + i\left(\frac{2}{\pi\zeta}\right)^{1/2}\cos\zeta\right]$$

$$= i\left(\frac{2}{\pi}\right)^{1/2}A\,e^{-i\zeta}$$

showing that there is a wave propagating only in the $+z$ direction, which should indeed be the case. The variation of $|Z(\zeta)|^2$ with ζ for $p = 0$ and $p = \frac{1}{3}$ has been given in Fig. 8.1. It should be noted that there is no mode conversion; and that at an arbitrary value of z

$$\psi(x, y, z) = \sum_m \sum_n A_{mn} X_n(x) Y_m(y) Z_{mn}(\zeta) \tag{8.13}$$

where, in this particular case, $X_n(x)$ and $Y_m(y)$ are Hermite–Gauss functions

$$X_n(x) = N_n H_n(x/x_0) \exp[-\tfrac{1}{2}(x/x_0)^2]$$
$$Y_m(y) = N_m H_m(y/y_0) \exp[-\tfrac{1}{2}(y/y_0)^2] \tag{8.14}$$

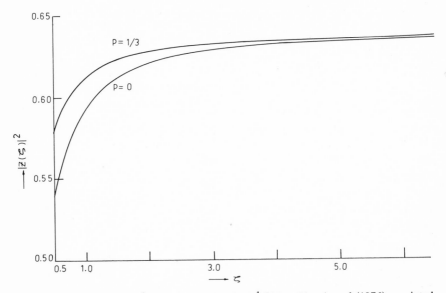

Fig. 8.1. Variation of $|Z(\zeta)|^2$ with ζ for $p = 0$ and $p = \frac{1}{3}$. [After Ghatak *et al.* (1976); reprinted by permission.]

where

$$x_0 = (c/\omega K_2^{1/2})^{1/2}$$
$$y_0 = (c/\omega K_3^{1/2})^{1/2}$$

and N_n and N_m are normalization constants.

We conclude this section by noting that if the z dependence of the dielectric constant can be separated out then there is no mode conversion; however, the amplitude of the mode will vary with z.

The shape of a temporal pulse [corresponding to the (n, m)th mode] at the output end of the fiber would be given by [see Eq. (7.34)]

$$q_{nm}(t) = \frac{1}{(2\pi)^{1/2}} \int_{-\infty}^{+\infty} F(\omega) S_{nm}(\omega) \exp(i\omega t)\, d\omega \qquad (8.15)$$

where the amplitude transfer function, $S_{nm}(\omega)$, is simply Z_{nm}, which is given by Eq. (8.11). Further,

$$F(\omega) = \frac{1}{(2\pi)^{1/2}} \int f(t) \exp(-i\omega t)\, dt \qquad (8.16)$$

where $f(t)$ represents the time dependence of the incident pulse, which for a Gaussian temporal distribution is given by

$$f(t) = \exp[-(t^2/2t_i^2) + i\omega_c t] \qquad (8.17)$$

for which

$$F(\omega) = t_i \exp\left[-\frac{(\omega - \omega_c)^2 t_i^2}{2} \right] \qquad (8.18)$$

8.2. $K = K_0 - K_2(z)r^2$

We next consider the case when the dielectric constant profile is given by

$$K = K_0 - K_2(z)r^2 \qquad (8.19)$$

We again start with the scalar wave equation

$$\nabla^2 \psi + \frac{\omega^2}{c^2}[K_0 - K_2(z)r^2]\psi = 0 \qquad (8.20)$$

and try to obtain a direct solution. The method is similar to the one given by Sodha et al. (1971a, b)* except that we wish to obtain an expression for the

*See also Chapter 5.

field ψ rather than the intensity. We assume an azimuthally symmetric solution of the form

$$\psi = A(r, z) \exp[i(\omega t - kz)] \tag{8.21}$$

to obtain

$$2ik\frac{\partial A}{\partial z} = \frac{1}{r}\frac{\partial}{\partial r}\left(r\frac{\partial A}{\partial r}\right) - \frac{k^2 K_2(z)}{K_0}r^2 A$$

where $k = (\omega/c)K_0^{1/2}$ and we have assumed $\partial^2 A/\partial z^2$ to be negligible. We write

$$A = A_0(r, z) \exp[-ikS(r, z)] \tag{8.22}$$

to obtain

$$2\frac{\partial S}{\partial z} + \left(\frac{\partial S}{\partial r}\right)^2 = -\frac{K_2(z)}{K_0}r^2 + \frac{1}{k^2 A_0}\left(\frac{\partial^2 A_0}{\partial r^2} + \frac{1}{r}\frac{\partial A_0}{\partial r}\right) \tag{8.23}$$

and

$$\frac{\partial A_0^2}{\partial z} + \frac{\partial S}{\partial r}\frac{\partial A_0^2}{\partial r} + A_0^2\left(\frac{\partial^2 S}{\partial r^2} + \frac{1}{r}\frac{\partial S}{\partial r}\right) = 0 \tag{8.24}$$

where use has been made of the fact that A_0 and S are real functions. Following Sodha *et al.* (1971a,b), if we assume a solution of the form

$$S(r, z) = (r^2/2)\beta(z) + \varphi(z) \tag{8.25}$$

then the general solution of Eq. (8.24) will be of the form

$$A_0^2 = \frac{E_0^2}{f^2(z)}\mathscr{F}\left[\frac{r}{af(z)}\right] \tag{8.26}$$

where*

$$\frac{1}{f}\frac{df}{dz} = \beta(z) \tag{8.27}$$

and $\mathscr{F}$ is an arbitrary function of its argument. If we assume an incident beam having a Gaussian intensity distribution, then

$$A_0^2 = \frac{E_0^2}{f^2(z)}\exp\left[-\frac{r^2}{a^2 f^2(z)}\right] \tag{8.28}$$

where $f(z) = 1$ at $z = 0$ and a represents the initial spot size of the beam. Substituting for A_0 and S in Eq. (8.23) and equating the coefficients of equal

*$[(1/f)(df/dz)]^{-1}$ represents the radius of curvature of the wavefront.

powers of r^2, we obtain

$$\frac{1}{f}\frac{d^2f}{dz^2} = -\frac{K_2(z)}{K_0} + \frac{1}{k^2a^4f^4} \tag{8.29}$$

and

$$\frac{d\varphi}{dz} = -\frac{2}{k^2a^2f^2} \tag{8.30}$$

Thus the complete solution will be of the form

$$\psi = \frac{E_0}{f(z)}\exp\left[-\frac{r^2}{2a^2f^2(z)}\right]\exp\left(i\left\{\omega t - k\left[z + \frac{r^2}{2}\frac{1}{f}\frac{df}{dz} + \varphi(z)\right]\right\}\right) \tag{8.31}$$

We expand this in terms of the modes for a perfect fiber for which

$$K_2(z) = K_{20} \tag{8.32}$$

Thus, we write

$$\psi(r, z) = \sum_{n=0}^{\infty} A_n(z)u_n(r) \tag{8.33}$$

where $u_n(r)$ are the normalized Laguerre Gauss functions which are the only modes excited for an azimuthally symmetric profile (Gambling and Matsumura, 1973; see also Chapter 7) and

$$u_n(r) = \frac{1}{\sqrt{\pi}w_0}L_n\left(\frac{r^2}{w_0^2}\right)\exp\left(-\frac{r^2}{2w_0^2}\right) \tag{8.34}$$

where

$$w_0 = \left(\frac{K_0}{K_{20}k^2}\right)^{1/4}$$

Using the orthonormal properties of $u_n(r)$, we obtain

$$A_n(z) = \frac{2\sqrt{\pi}}{w_0}\int_0^{\infty} \psi(r, z)L_n\left(\frac{r^2}{w_0^2}\right)\exp\left(-\frac{r^2}{2w_0^2}\right)r\,dr$$

$$\tag{8.36}$$

$$= \frac{2\sqrt{\pi}E_0w_0}{f(z)}\exp\{i[\omega t - kz - k\varphi(z)]\}\frac{1}{1+W^2(z)}\left(\frac{W^2-1}{W^2+1}\right)^n$$

where

$$W(z) = \frac{w_0}{w_i(z)}$$

$$\tag{8.37}$$

$$w_i(z) = \left(\frac{1}{a^2f^2} + \frac{ik}{f}\frac{df}{dz}\right)^{-1/2}$$

Thus

$$W^2 \pm 1 = \left(\frac{w_0^2}{a^2 f^2} \pm 1\right) + \frac{ikw_0^2}{f}\frac{df}{dz}$$

and

$$|W^2 \pm 1| = \left[\left(\frac{w_0^2}{a^2 f^2} \pm 1\right)^2 + \frac{k^2 w_0^4}{f^2}\left(\frac{df}{dz}\right)^2\right]^{1/2}$$

Hence the z dependence of the power associated with the nth mode would be proportional to

$$|A_n|^2 = \frac{4\pi E_0^2 w_0^2}{f^2(z)} \; \frac{\left[\left(\frac{w_0^2}{a^2 f^2} - 1\right)^2 + k^2 w_0^4\left(\frac{1}{f}\frac{df}{dz}\right)^2\right]^n}{\left[\left(\frac{w_0^2}{a^2 f^2} + 1\right)^2 + k^2 w_0^4\left(\frac{1}{f}\frac{df}{dz}\right)^2\right]^{(n+1)}} \tag{8.38}$$

For a perfect waveguide [see Eq. (8.32)] a straightforward integration of Eq. (8.29) gives

$$\left(\frac{1}{f}\frac{df}{dz}\right)^2 = -\frac{K_2}{K_0} + \frac{K_2}{K_0 f^2} - \frac{1}{k^2 a^4 f^4} + \frac{1}{k^2 a^4 f^2} \tag{8.39}$$

where we have used the condition that at $z = 0$, $f = 1$, and $df/dz = 0$. It can easily be shown that

$$|W^2(z) \pm 1| = \frac{1}{f(z)}\left|1 \pm \frac{W_0^2}{a^2}\right| \tag{8.40}$$

On substitution in Eq. (8.36), we get $|A_n|^2$ to be independent of z, showing no mode conversion. When K_2 depends on z one should solve Eq. (8.29) to obtain explicit z dependence of f and df/dz. Once this is known one can calculate the z dependence of $|A_n|^2$ and hence the power excited in other modes due to mode conversion. For example, if the variation of $K_2(z)$ is of the form

$$K_2(z) = \frac{K_{20}}{(1 + \alpha z)^2} \tag{8.41}$$

then an analytical solution of Eq. (8.29) can be obtained (see Chapter 5):

$$f^2(z) = (1 + \alpha z)\left\{\left[\frac{1}{2} + \frac{1}{\gamma^2}\left(\frac{1}{8} + \frac{\Lambda}{2}\right)\right] - \frac{1}{2\gamma}\sin[2\gamma \ln(1 + \alpha z)]\right.$$

$$\left. + \left[\frac{1}{2} - \frac{1}{\gamma^2}\left(\frac{1}{8} + \frac{\Lambda}{2}\right)\right]\cos[2\gamma \ln(1 + \alpha z)]\right\} \tag{8.42}$$

where

$$\Lambda = \frac{1}{w_0^4 \alpha^2 k^2}$$

$$\gamma = \left(\frac{1}{\beta^2} - \frac{1}{4}\right)^{1/2}$$

$$\beta = \alpha \left(\frac{K_0}{K_{20}}\right)^{1/2}$$

and other symbols have been defined in Chapter 5. [The solution given by Eq. (8.42) is valid for $\beta < 2$ which is indeed the case for a realistic fiber.] On substitution of the above expression for $f(z)$ in Eq. (8.38) we get the variation of the power (in a particular mode) with z due to mode conversion.

In Fig. 8.2, we have plotted a typical variation of f^2 with z corresponding to $\alpha = 10^{-5}$ cm^{-1} and $K_{20}/K_0 = 2500$ cm^{-2}. The steady increase in the value of f^2 can be understood by noting that if z is such that

$$2\gamma \ln(1 + \alpha z) = 2m\pi \qquad m = 0, 1, 2, \ldots$$

then

$$f^2(z) = (1 + \alpha z)$$

There are rapid oscillations of $f^2(z)$ which are superimposed on this steady increase (see inset in Fig. 8.2).

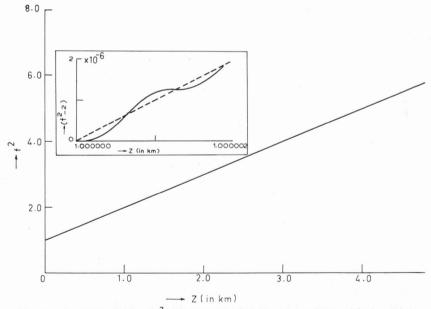

Fig. 8.2. A typical variation of f^2 with z for a dielectric constant variation of the form given by Eq. (8.41). Superimposed on the smooth increase is an oscillatory variation which has been shown in the inset. [After Ghatak et al. (1976), reprinted by permission.]

In order to study mode conversion, we have plotted in Fig. 8.3 the variation of $|A_0|^2$ and $|A_1|^2$ with z. We have assumed that only the fundamental mode is excited at the entrance aperture of the guide ($z = 0$). These functions undergo rapid oscillations about a mean curve; this mean curve can be obtained by calculating $|A_n|^2$ for those values of z for which

$$2\gamma \ln(1+\alpha z) = 2m\pi \qquad m = 0, 1, 2, \ldots$$

At these values

$$|A_0|^2 = \text{sech}^2[\tfrac{1}{2}\ln(1+\alpha z)] = \left[1 - \frac{(\alpha z)^2}{(1+\alpha z)^2}\right]$$

and

$$|A_n|^2 = \text{sech}^2[\tfrac{1}{2}\ln(1+\alpha z)]\,\tanh^{2n}[\tfrac{1}{2}\ln(1+\alpha z)]$$

The variation of $|A_n|^2$ with z shows mode conversion.

In general, for an arbitrary z dependence of K_2, one has to carry out numerical solutions of Eq. (8.19) which are not very difficult to perform. Such numerical solutions have been obtained by Sodha *et al.* (1971b).

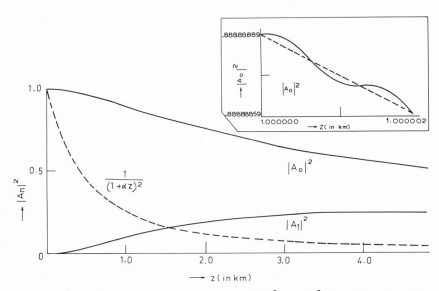

Fig. 8.3. The solid curves represent the variation of $|A_0|^2$ and $|A_1|^2$ with z. Superimposed on the smooth curve is an oscillatory variation which has been shown in the inset. The corresponding variation of $K_2(z)$ is shown as a dashed curve. [After Ghatak *et al.* (1976), reprinted by permission.]

8.3. $K = K^{(0)}(x, y) + K^{(p)}(x, y, z)$

In this section we will develop the Green's function method to study the mode conversion due to irregularities in a waveguide. The dielectric constant variation will be assumed to be of the form

$$K = K^{(0)}(x, y) + K^{(p)}(x, y, z) \qquad (8.43)$$

where $K^{(0)}(x, y)$ corresponds to the perfect waveguide and $K^{(p)}(x, y, z)$ represents the irregularity which is responsible for mode conversion. The superscript p refers to the fact that $K^{(p)}(x, y, z)$ is to be treated as a perturbation. The scalar wave equation is written in the form

$$\nabla^2 \psi + (\omega^2/c^2) K^{(0)}(x, y)\psi = F(\mathbf{r}) \qquad (8.44)$$

where

$$F(\mathbf{r}) = -(\omega^2/c^2) K^{(p)}(x, y, z)\psi(x, y, z) \qquad (8.45)$$

Equation (8.44) may be rewritten in the form

$$\left(\frac{\partial^2}{\partial z^2} + L\right)\psi = F(\mathbf{r}) \qquad (8.46)$$

where the operator L is defined by the following equation

$$L = \frac{\partial^2}{\partial x^2} + \frac{\partial^2}{\partial y^2} + K^{(0)}(x, y)\frac{\omega^2}{c^2} \qquad (8.47)$$

In order to solve the inhomogeneous equation [Eq. (8.46)], we write*

$$\psi(x, y, z) = \sum_n \int_{-\infty}^{+\infty} d\Lambda \, A(n, \Lambda) u_n(x, y) \exp(-i\Lambda z)$$

$$+ \int_{-\infty}^{+\infty} d\Lambda \int d\beta \, A(\beta, \Lambda) u_\beta(x, y) \exp(-i\Lambda z) \qquad (8.48)$$

where $\exp(-i\Lambda z)$ are the eigenfunctions of the operator $\partial^2/\partial z^2$:

$$\frac{\partial^2}{\partial z^2} \exp(-i\Lambda z) = (-\Lambda^2) \exp(-i\Lambda z) \qquad (8.49)$$

---

*In the second term on the RHS of Eq. (8.48), the limits on the integral over β are determined from the characteristics of the waveguide. For example, for the waveguide characterized by Eq. (8.52a) the limits are from $-(\omega/c)K_2^{1/2}$ to $+(\omega/c)K_2^{1/2}$ [see Eq. (8.52c)].

and $u_n(x, y)$ and $u_\beta(x, y)$ represent the discrete and continuum eigenfunctions of the operator L:

$$Lu_n(x, y) = \beta_n^2 u_n(x, y)$$
$$Lu_\beta(x, y) = \beta^2 u_\beta(x, y)$$

(8.50)

β_n^2 and β^2 being the corresponding eigenvalues. [In Eq. (8.48) we have a sum over the discrete modes and an integration over the continuum.] For example, for a square-law medium

$$K^0(x, y) = K_0 - K_2(x^2 + y^2)$$

(8.51a)

we only have a discrete spectrum, the propagation constants of which are given by [see Eqs. (8.5) and (8.6)]:

$$\beta_{nm} = [K_0(\omega^2/c^2) - (2n + 2m + 2)(\omega/c)K_2^{1/2}]^{1/2}$$

(8.51b)

$$n, m = 0, 1, 2, \ldots$$

the corresponding eigenfunctions being Hermite–Gauss functions. On the other hand, for a cladded fiber of the type

$$K = K_1 \qquad 0 < r < a$$
$$= K_2 \qquad r > a$$

(8.52a)

(see Chapter 4), the operator will have discrete spectrum with

$$\frac{\omega^2}{c^2} K_2 < \beta^2 < \frac{\omega^2}{c^2} K_1$$

(8.52b)

and a continuum of modes for which

$$\beta^2 < (\omega^2/c^2)K_2$$

(8.52c)

The discrete modes are known as guided modes, whereas the continuum modes are known as radiation modes.

We substitute for ψ from Eq. (8.48) in Eq. (8.46) and use Eqs. (8.49) and (8.50) to obtain

$$\sum_n \int d\Lambda \, A(n, \Lambda)(\beta_n^2 - \Lambda^2)u_n \exp(-i\Lambda z)$$
$$+ \int d\Lambda \int d\beta \, A(\beta, \Lambda)(\beta^2 - \Lambda^2)u_\beta \exp(-i\Lambda z) = F(\mathbf{r})$$

(8.53)

If we multiply Eq. (8.53) by $(1/2\pi) \exp(i\Lambda' z)$ and integrate from $-\infty$ to $+\infty$ and make use of the relation

$$\frac{1}{2\pi} \int_{-\infty}^{+\infty} \exp[i(\Lambda' - \Lambda)z] \, dz = \delta(\Lambda - \Lambda')$$

(8.54)

we would obtain

$$\sum_n \int A(n, \Lambda)(\beta_n^2 - \Lambda^2)u_n(x, y)\delta(\Lambda - \Lambda')\, d\Lambda$$

$$+ \iint A(\beta, \Lambda)(\beta^2 - \Lambda^2)u_\beta(x, y)\delta(\Lambda - \Lambda')\, d\Lambda\, d\beta$$

$$= \frac{1}{2\pi} \int F(\mathbf{r}) \exp(i\Lambda' z)\, dz$$

Carrying out the integration over β and dropping the primes we get

$$\sum_n A(n, \Lambda)(\beta_n^2 - \Lambda^2)u_n(x, y) + \int A(\beta, \Lambda)(\beta^2 - \Lambda^2)u_\beta(x, y)\, d\beta$$

$$= \frac{1}{2\pi} \int F(\mathbf{r}) \exp(i\Lambda z)\, dz \qquad (8.55)$$

We multiply the above equation by $u_{n'}^*(x, y)$ and integrate over the transverse cross section of the waveguide. Using the orthonormal property of the wave functions

$$\iint u_{n'}^*(x, y)u_n(x, y)\, dx\, dy = \delta_{nn'} \qquad (8.56)$$

we get,

$$\sum A(n, \Lambda)(\beta_n^2 - \Lambda^2)\delta_{nn'} = \frac{1}{2\pi} \iiint F(x, y, z)u_{n'}^*(x, y) \exp(i\Lambda z)\, dx\, dy\, dz \qquad (8.57)$$

In Eq. (8.56) $\delta_{nn'}$ represents the Kronecter delta function. Carrying out the summation and dropping the primes we obtain

$$A(n, \Lambda) = \frac{1}{2\pi(\beta_n^2 - \Lambda^2)} \iiint F(\mathbf{r})u_n^*(x, y) \exp(i\Lambda z)\, dx\, dy\, dz \qquad (8.58)$$

Similarly*

$$A(\beta, \Lambda) = \frac{1}{2\pi(\beta^2 - \Lambda^2)} \iiint F(\mathbf{r})u_\beta^*(x, y) \exp(i\Lambda z)\, dx\, dy\, dz \qquad (8.59)$$

Thus Eq. (8.48) can be written in the form

$$\psi = \iiint F(\mathbf{r'})G(\mathbf{r}, \mathbf{r'})\, d\mathbf{r'} \qquad (8.60)$$

*The continuum eigenfunctions satisfy the following orthonormality relation $\iint u_{\beta'}^*(x, y)u_\beta(x, y)\, dx\, dy = \delta(\beta - \beta')$ where $\delta(\beta - \beta')$ is the Dirac delta function.

which represents the solution to the inhomogeneous part of Eq. (8.46); the Green's function G is given by

$$G(\mathbf{r}, \mathbf{r}') = \frac{1}{2\pi} \sum_n \int \frac{u_n^*(x', y') u_n(x, y) \exp[-i\Lambda(z - z')]}{\beta_n^2 - \Lambda^2} d\Lambda$$

$$+ \frac{1}{2\pi} \iint \frac{u_\beta^*(x', y') u_\beta(x, y) \exp[-i\Lambda(z - z')]}{\beta^2 - \Lambda^2} d\beta \, d\Lambda \qquad (8.61)$$

We next consider the integral

$$I = \int_{-\infty}^{+\infty} \frac{\exp[-i\Lambda(z - z')]}{\beta^2 - \Lambda^2} d\Lambda \qquad (8.62)$$

The integrand has simple poles at $\Lambda = \pm\beta$. The evaluation of the integral is along the same lines as those followed in quantum mechanics (see, for example, Schiff, 1955, Sec. 26). For $z' > z$ we choose* a contour of the type shown in Fig. 8.4a. The contribution from the semicircle vanishes as $R \to \infty$ (Jordan's lemma) and

$$I = 2\pi i \times [\text{residue at } \Lambda = -\beta]$$

$$= 2\pi i \left[\frac{\exp(-i\beta|z - z'|)}{2\beta} \right]$$

$$= \frac{\pi i}{\beta} \exp(-i\beta|z - z'|) \qquad (8.63)$$

Similarly for $z > z'$, we choose a contour of the type shown in Fig. 8.4b and one obtains the same expression for I. If we substitute the above expressions in Eq. (8.61) we obtain

$$G = \frac{i}{2} \left[\sum_n \frac{1}{\beta_n} u_n^*(x', y') u_n(x, y) \exp\left(-i\beta_n|z - z'|\right) \right.$$

$$\left. + \int \frac{1}{\beta} u_\beta^*(x', y') u_\beta(x, y) \exp\left(-i\beta|z - z'|\right) d\beta \right] \qquad (8.64)$$

On substituting the above expression for G in Eq. (8.60) we get

$$\psi = \iiint F(\mathbf{r}') G(\mathbf{r}, \mathbf{r}') \, d\mathbf{r}'$$

$$= \frac{\omega^2}{2ic^2} \iiint \left[\sum_n \frac{1}{\beta_n} u_n^*(x', y') u_n(x, y) \exp\left(-i\beta_n|z - z'|\right) \right.$$

$$\left. + \int \frac{1}{\beta} u_\beta^*(x', y') u_\beta(x, y) \exp\left(-i\beta|z - z'|\right) d\beta \right]$$

$$\times K^{(p)}(x', y', z') \psi(x', y', z') \, dx' \, dy' \, dz' \qquad (8.65)$$

*The choice of the contour is determined by the boundary conditions.

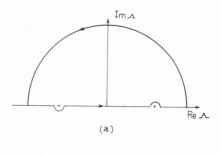

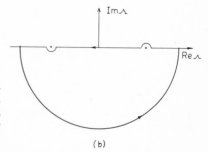

Fig. 8.4. The contours of integration for evaluating the integral in Eq. (8.62); (a) and (b) correspond to $z' > z$ and $z > z'$, respectively. The two crosses on the real axis represent the position of the poles of the integrand.

This completes the formulation of the problem. However, it should be noted that ψ appears in both sides of the equation; thus we have not really solved for ψ. We therefore make the approximation that on the RHS of Eq. (8.65) we substitute for ψ the field pattern of the incident mode and assume that mode conversion is small,* which essentially implies that $K^{(p)}(x, y, z)$ is treated as a perturbation.

8.4. Application

As an application of the above results we consider a length L of an imperfect waveguide $[K^{(p)}(x, y, z) \neq 0]$ and assume that beyond this length there are no imperfections, i.e.,

$$K^{(p)}(x, y, z) = 0 \qquad \text{for } z > L$$

Assuming that at $z = 0$ only the fundamental mode is excited, we will calculate the power converted into other modes after the beam has propagated through a length L of the imperfect waveguide. In carrying out the analysis we will assume that as the beam propagates through the waveguide,

_____

*We will show later that the treatment will be valid for appreciable power getting converted to radiation modes.

most of the power is associated with the incident mode; consequently we will assume

$$F(x, y, z) \approx -K^{(p)}(x, y, z)(\omega^2/c^2)u_0(x, y) \exp(-i\beta_0 z) \qquad (8.66)$$

where the subscript zero refers to the fundamental mode. Since $K^{(p)} = 0$ for $z > L$, we have [see Eq. (8.60)]

$$\psi = -(\omega^2/c^2) \int_0^L dz' \int_{-\infty}^{+\infty} dx' \int_{-\infty}^{\infty} dy' \, K^{(p)}(x', y', z')u_0(x', y')$$

$$\times \exp(-i\beta_0 z')G(\mathbf{r}, \mathbf{r}') \qquad (8.67)$$

If we use Eq. (8.64) for $G(\mathbf{r}, \mathbf{r}')$, Eq. (8.67) can be written in the form

$$\psi = \sum_n A_n(L)u_n(x, y) \exp(-i\beta_n z) + \int A_\beta(L)u_\beta(x, y) \exp(-i\beta z) \, d\beta \qquad (8.68)$$

where

$$A_n(L) = \frac{\omega^2}{2ic^2\beta_n} \int_0^L dz' \int_{-\infty}^{\infty} dx' \int_{-\infty}^{+\infty} dy' \, u_n^*(x', y')u_0(x', y')$$

$$\times K^{(p)}(x', y', z') \exp[-i(\beta_0 - \beta_n)z'] \qquad (8.69a)$$

and

$$A_\beta(L) = \frac{\omega^2}{2ic^2\beta} \int_0^L dz' \int_{-\infty}^{+\infty} dy' \int_{-\infty}^{+\infty} dx' \, u_\beta^*(x', y')u_0(x', y')$$

$$\times K^{(p)}(x', y', z') \exp[-i(\beta_0 - \beta)z'] \qquad (8.69b)$$

Notice that in writing the above equations we have assumed $z > z'$.

In order to carry out a specific application we consider a slab waveguide whose dielectric constant variation is of the form [see Eq. (2.41)]

$$\begin{aligned} K &= K_1 \qquad \text{for } |x| < a \\ &= K_2 \qquad \text{for } |x| > a \end{aligned} \qquad (8.70)$$

The patterns of TE modes (which in this case correspond to the y component of the electric field) are given by the following expressions (see Sec. 2.3).

8.4.1.   Guided Modes $[(\omega^2/c^2)K_2 < \beta_n^2 < (\omega^2/c^2)K_1]$

Symmetric Modes

$$\begin{aligned} u_{ns} &= [a + (1/\kappa_n)]^{-1/2} \cos p_n x & \text{for } |x| \leq a \\ &= [a + (1/\kappa_n)]^{-1/2} \cos p_n a \, \exp[-\kappa_n(|x| - a)] & \text{for } |x| \geq a \end{aligned} \qquad (8.71)$$

Antisymmetric Modes

$$u_{na} = [a + (1/\kappa_n)]^{-1/2} \sin p_n x \qquad \text{for } |x| \leq a$$
$$= [a + (1/\kappa_n)]^{-1/2} \sin p_n a \, \exp[-\kappa_n(|x| - a)] \qquad \text{for } |x| \geq a \qquad (8.72)$$

where

$$p_n^2 = (\omega^2/c^2)K_1 - \beta_n^2$$
$$\kappa_n^2 = \beta_n^2 - (\omega^2/c^2)K_2 \qquad (8.73)$$

For guided modes the propagation constants are determined from the following equation

$$\tan pa = \pm \frac{p}{\kappa} \qquad (8.74)$$

where the $+$ and $-$ sign correspond to symmetric and antisymmetric modes, respectively.

8.4.2. Radiation Modes

The radiation modes form a continuum and their propagation constants satisfy the inequality

$$\beta^2 < (\omega^2/c^2)K_2 \qquad (8.75)$$

The field patterns are given below*:

Symmetric Modes

$$u_{\beta s} = C_s \cos \sigma x \qquad \text{for } |x| < a$$
$$= D_s \exp(-i\rho|x|) + D_s^* \exp(i\rho|x|) \qquad \text{for } |x| > a \qquad (8.76)$$

where

$$D_s = \tfrac{1}{2} \exp(i\rho a)[\cos \sigma a - (i\sigma/\rho) \sin \sigma a]C_s$$
$$C_s = \left[\frac{\rho^2}{\pi(\rho^2 \cos^2 \sigma a + \sigma^2 \sin^2 \sigma a)}\right]^{1/2}\left(\frac{|\beta|}{\rho}\right)^{1/2}$$
$$\sigma = [(\omega^2/c^2)K_1 - \beta^2]^{1/2}$$
$$\rho = [(\omega^2/c^2)K_2 - \beta^2]^{1/2} \qquad (8.77)$$

*The notation is the same as in Marcuse (1972). The field patterns can readily be obtained by assuming the continuity of E_y and H_z at $x = \pm a$ (see Chapter 2).

Antisymmetric Modes

$$u_{\beta a} = C_a \sin \sigma x \qquad\qquad\qquad \text{for } |x| < a$$
$$= (x/|x|)[D_a \exp(-i\rho|x|) + D_a^* \exp(i\rho|x|)] \qquad \text{for } |x| > a \tag{8.78}$$

where

$$D_a = \tfrac{1}{2} \exp(i\rho a) \left[\sin \sigma a + (i\sigma/\rho) \cos \sigma a\right] C_a$$

and

$$C_a = \left[\frac{\rho^2}{\pi(\rho^2 \sin^2 \sigma a + \sigma^2 \cos^2 \sigma a)}\right]^{1/2} \left(\frac{|\beta|}{\rho}\right)^{1/2}$$

The normalization constants in Eqs. (8.71), (8.72), (8.76), and (8.78) are such that

$$\int_{-\infty}^{+\infty} u_{n'}^*(x) u_n(x)\, dx = \delta_{n,n'} \tag{8.79}$$

and

$$\int_{-\infty}^{+\infty} u_{\beta'}^*(x) u_\beta(x)\, dx = \delta(\beta - \beta') \tag{8.80}$$

Further, the power associated with such normalized modes would be $\beta/\omega\mu_0$. As a model for the imperfection in such a waveguide, we assume that the interface is *not* always at a distance a from the axis. Such a model was discussed by Marcuse (1969) and the variation of $K^{(0)}$ and $K^{(p)}$ [see Eq. (8.43)] is assumed to be of the form (see Fig. 8.5)

$$K^{(0)} = K_2 \qquad\qquad \text{for } |x| > a$$
$$= K_1 \qquad\qquad \text{for } |x| < a \tag{8.81}$$

and

$$K^{(p)} = 0 \qquad\qquad\qquad \text{for } x > a + f_1(z)$$
$$= K_1 - K_2 \qquad\qquad \text{for } a < x < a + f_1(z)$$
$$= 0 \qquad\qquad\qquad \text{for } -a + f_2(z) < x < a \tag{8.82}$$
$$= -(K_1 - K_2) \qquad\quad \text{for } -a < x < -a + f_2(z)$$
$$= 0 \qquad\qquad\qquad \text{for } x < -a$$

Thus at an arbitrary value of z the interface occurs at

$$x = a + f_1(z) \tag{8.83a}$$

and at

$$x = -a + f_2(z) \tag{8.83b}$$

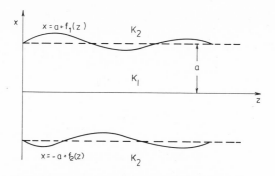

Fig. 8.5. The dashed lines represent the core-cladding interface for a perfect slab waveguide. The solid curves denote core–cladding interface due to imperfections in the waveguide (after D. Marcuse, Light Transmission Optics, Van Nostrand Reinhold Company, reproduced by permission).

instead of $x = \pm a$. Obviously, for the perfect waveguide $f_1 = f_2 = 0$. We assume that

$$f_1(z), f_2(z) \ll a \qquad (8.84)$$

which will indeed be the case for an actual waveguide. Consequently, Eq. (8.69a) can be written as

$$A_n(L) = \frac{\omega^2}{2ic^2\beta_n} \int_0^L dz' \left[-(K_1 - K_2) \int_{-a}^{-a+f_2} u_n^*(x')u_0(x')\,dx' \right.$$

$$\left. + (K_1 - K_2) \int_a^{a+f_1} u_n^*(x')u_0(x')\,dx' \right] \exp[-i(\beta_0 - \beta_n)z'] \qquad (8.85)$$

Notice that because of the slab geometry, there is no integration over the y coordinate. Since f_1 and f_2 are very small in comparison to a we may assume that the functions $u_n^*(x')$ and $u_0(x')$ remain constant in the region of integration (over x'). Thus we may write

$$A_n(L) \approx \frac{\omega^2}{2ic^2\beta_n}(K_1 - K_2) \int_0^L dz' [u_n^*(a)u_0(a)f_1(z')$$

$$- u_n^*(-a)u_0(-a)f_2(z')] \exp[-i(\beta_0 - \beta_n)z'] \qquad (8.86)$$

If we are considering the excitation of a symmetric mode,[*] then

$$u_n^*(a) = u_n^*(-a)$$

[*]The fundamental mode will be, of course, a symmetric mode. If we assume any other mode to be initially excited, the treatment will accordingly be modified.

and Eq. (8.86) simplifies to

$$A_{ns}(L) \approx \frac{\omega^2}{2ic^2\beta_n}(K_1 - K_2)u_{ns}(a)u_0(a)L(F_{1n} - F_{2n}) \qquad (8.87)$$

where

$$F_{in} \equiv \frac{1}{L}\int_0^L f_i(z)\exp[-i(\beta_0 - \beta_n)z]\,dz \qquad (8.88)$$

Similarly, if the nth mode is antisymmetric, then

$$A_{na}(L) \approx \frac{\omega^2}{2ic^2\beta_n}(K_1 - K_2)u_{na}(a)u_0(a)L(F_{1n} + F_{2n}) \qquad (8.89)$$

In Eqs. (8.87) and (8.89), the subscripts s and a refer to symmetric and antisymmetric modes, respectively. Similarly, we may consider the excitation of the continuum modes and the final results are as follows:

$$A_{\beta s}(L) \approx \frac{\omega^2}{2ic^2\beta}(K_1 - K_2)u_{\beta s}(a)u_0(a)L(F_{1n} - F_{2n}) \qquad (8.90)$$

and

$$A_{\beta a}(L) = \frac{\omega^2}{2ic^2\beta}(K_1 - K_2)u_{\beta a}(a)u_0(a)(F_{1n} + F_{2n}) \qquad (8.91)$$

Following Marcuse (1969), we assume

$$f_1(z) = \varepsilon\sin\theta z \qquad (8.92a)$$

and

$$f_2(z) = -\varepsilon\sin(\theta z + \phi) \qquad (8.92b)$$

If $\phi = 0$ then the width of the core will vary periodically as shown in Fig. 8.6a. On the other hand if $\phi = \pi$, the core–cladding interface will be as shown in Fig. 8.6b. Now, if we substitute for $f_1(z)$ in Eq. (8.88) we would get

$$F_{1n} = \frac{1}{L}\varepsilon\int_0^L\left[\frac{\exp(i\theta z) - \exp(-i\theta z)}{2i}\right]\exp(-i\gamma_n z) \qquad (8.93)$$

where

$$\gamma_n = \beta_0 - \beta_n$$

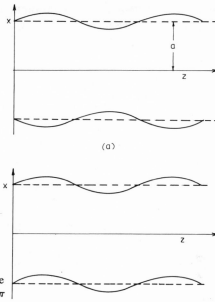

Fig. 8.6. The sinusoidal variation of the core cladding interface with $\phi = 0$ (a) and $\phi = \pi$ (b).

Thus

$$F_{1n} = \frac{\varepsilon}{2iL}\left\{\frac{\exp[i(\theta - \gamma_n)z]}{i(\theta - \gamma_n)} + \frac{\exp[-i(\theta + \gamma_n)z]}{i(\theta + \gamma_n)}\right\}_0^L$$

$$= \frac{\varepsilon}{2iL}\left\{\exp[i(\theta - \gamma_n)L/2]\frac{\sin[(\theta - \gamma_n)L/2]}{[(\theta - \gamma_n)/2]}\right.$$

$$\left. - \exp[-i(\theta + \gamma_n)L/2]\frac{\sin[(\theta + \gamma_n)L/2]}{[(\theta + \gamma_n)/2]}\right\} \qquad (8.94)$$

It can easily be seen that for large values of L the function

$$\frac{\sin Lx}{x}$$

is very sharply peaked about $x = 0$. In fact

$$\lim_{L \to \infty} \frac{\sin Lx}{x} = \pi\delta(x) \qquad (8.95)$$

Consequently, if the imperfection has the periodic form of the type given by Eq. (8.92) then only those modes for which

$$\theta = \gamma_n = \beta_0 - \beta_n \qquad (8.96)$$

will be excited. The above condition holds for guided modes as well as for radiation modes. Thus, for $\phi = 0$, Eq. (8.94) simplifies to

$$F_{1n} = 0$$
$$\qquad\qquad n \neq m \qquad\qquad\qquad (8.97)$$
$$F_{2n} = 0$$

$$F_{1m} \approx \frac{\varepsilon}{2i} \qquad\qquad\qquad\qquad (8.98a)$$

$$F_{2m} \approx -\frac{\varepsilon}{2i} \qquad\qquad\qquad\qquad (8.98b)$$

where m refers to that mode for which

$$\beta_o - \beta_m = \theta \qquad\qquad\qquad\qquad (8.99)$$

Since $F_{1m} + F_{2m} = 0$, the antisymmetric modes are not excited. Had we used $\phi = \pi$, the sign of F_2 [see Eq. (8.98b)] would have been positive and coupling could occur only to antisymmetric modes. Now, the power associated with a particular mode is

$$(\beta/\omega\mu_0)|A_n|^2 \qquad\qquad\qquad\qquad (8.100)$$

Thus, the relative power loss of the incident mode to the mth mode (to which it is coupled) would be given by

$$\left(\frac{\Delta P}{P}\right)_g = \frac{\beta_m |A_m(L)|^2}{\beta_0 |A_0(0)|^2}$$
$$= \varepsilon^2 L^2 \frac{[(K_1 - K_2)(\omega^2/2c^2)\cos p_m a \cos p_0 a]^2}{[a + (1/\kappa_m)][a + (1/\kappa_0)]\beta_0\beta_m} \qquad (8.101)$$

where we have assumed $\phi = 0$ and used Eqs. (8.71), (8.87), and (8.99). Similarly, one may consider the excitation of radiation modes for which

$$\beta^2 < (\omega^2/c^2)K_2$$

or

$$-(\omega/c)K_2^{1/2} < \beta < +(\omega/c)K_2^{1/2} \qquad\qquad (8.102)$$

Consequently, if we use Eq. (8.99) then for radiation modes to be excited θ has to lie between the following two limits

$$\beta_0 - (\omega/c)K_2^{1/2} < \theta < \beta_0 + (\omega/c)K_2^{1/2} \qquad\qquad (8.103)$$

The fractional power associated with radiation modes would be given by

$$\left(\frac{\Delta P}{P}\right)_r = \int_{-(\omega/c)K_2^{1/2}}^{+(\omega/c)K_2^{1/2}} \frac{\beta}{\beta_0}[|A_{\beta s}|^2 + |A_{\beta a}|^2]/d\beta$$

For $\phi = 0$ only the symmetric modes are excited, and if we substitute the expression for $A_{\beta s}$ in the above equation then the integrals can be evaluated and the final result is (see, for example, Marcuse, 1972):

$$\frac{\Delta P}{P} = 2\eta L \tag{8.104}$$

where

$$\eta = \frac{\varepsilon^2 (K_1 - K_2)^2 (\omega/c)^4 \rho_\theta \cos^2 p_0 a \cos^2 \sigma_\theta a}{2\beta_0 [a + (1/\kappa_0)](\rho_\theta^2 \cos^2 \sigma_\theta a + \sigma_\theta^2 \sin^2 \sigma_\theta a)}$$

with

$$\rho_\theta = [(\omega^2/c^2)K_2 - (\beta_0 - \theta)^2]^{1/2}$$
$$\sigma_\theta = [(\omega^2/c^2)K_1 - (\beta_0 - \theta)^2]^{1/2}$$

and use has been made of the fact that

$$\frac{1}{x} \sin\left(\frac{L}{2}x\right) \approx \pi\delta(x) \tag{8.105}$$

which will be valid for large values of L.

In the derivation of Eq. (8.101) and (8.104) we have assumed that most of the incident power continues to travel in the mode which is initially excited. In the case of coupling to a guided mode the power remains in the guide and may be converted back (partially) to the incident mode. Further, the power associated with the excited modes may become comparable to the power associated with the incident mode and the approximations made in the derivation of Eq. (8.101) will no longer be valid. However, in the case of radiation loss, the power associated with the radiation mode is radiated away from the system. Thus, if we divide the length of the waveguide into a large number of sections, then after the beam propagates through a section (of width Δz) the fractional power loss due to radiation will be given by

$$\frac{\Delta P}{P} = -2\eta \,\Delta z \tag{8.106}$$

where the negative sign is due to the fact that here ΔP represents the change in the power associated with the fundamental mode. On integration one simply obtains

$$P = P_0 \exp(-2\eta z) \tag{8.107}$$

which represents the exponential attenuation of the beam due to radiation losses.

Ray Tracing and Aberrations in Lens-like Media

The purpose of an optical instrument is to produce perfect images of objects, perfect in the sense that the image completely resembles the object. No single optical instrument is capable of producing exact images of every point of a three-dimensional region, i.e., a situation where all rays starting from any object point converge to a single image point. The departure of the image from perfectness is termed as aberration. In many of the applications of optical instruments one would like to know the aberrations present in the system.

In this chapter we will be calculating the aberrations introduced in rays as they travel through a rotationally symmetric system. We will be neglecting the effects due to diffraction, which is almost always justified whenever the dimensions of the system are large in comparison to the wavelength. Thus we will be dealing only with the ray theory of aberration and the term aberration used from now on will imply ray aberration. A detailed account of diffraction theory of aberration may be found in the books by Born and Wolf (1970) and Steward (1928).

The aberrations introduced by the system in rays traversing through it may be determined by either a ray tracing technique or by analytical methods. The ray tracing method involves step by step integration of the ray equation* corresponding to specific rays starting from the object point, and finding the intersection point in the image plane. The analytical method consists in obtaining the aberrations algebraically through equations which describe the propagation of rays through the system. The ray tracing method suffers from the limitation that one does not know the distribution of the aberration into various orders and further, the distribution of any order into the various kinds of aberration. The ray tracing technique also involves considerable computer calculations. It only gives the total aberration present in the system. On the other hand, the analytical method becomes more and more complicated as higher and higher orders of aberrations are treated, especially when media with continuously varying refractive index are also present in the system.

In the first part of the chapter we will deal with the analytical method using Hamiltonian optics and derive explicit expressions for the aberration. Sands (1970) and Moore (1971) have given a detailed account of the aberrations of cylindrically symmetric systems; they have used the theory of quasi-invariants as developed by Buchdahl (1968). This approach is different from the conventional Hamiltonian theory as given in detail by Luneburg (1964) and Buchdahl (1970). In this chapter we present the Hamiltonian theory of aberrations† as developed by Luneburg (1964). These studies involving media with continuously varying refractive index are relevant now, since rods with continuously decreasing refractive index have been experimentally prepared (see Appendix B) and these have short focal lengths and can be used in the form of microlenses.

In the second part we will deal with the ray tracing method and in the final part we will compare the two and show that for practical systems they do agree to a good extent. We would first outline the Hamiltonian formulation and use the formulas developed to derive explicit expressions for the aberration.

9.1. Hamiltonian Formulation

We specify our system with a refractive index variation given by $n(x, y, z)$. This would be a discontinuous function of position when the

---

*The ray equation is given by (see Appendix D)

$$\frac{d}{ds}\left(n\frac{d\mathbf{r}}{ds}\right) = \nabla n$$

where s represents an infinitesimal arc length along the ray, $\mathbf{r}$ represents the position vector of any point on the ray and $n(x, y, z)$ is the refractive index function.

†The Hamiltonian theory has been extended by Thyagarajan and Ghatak (1976) to include the presence of surfaces between inhomogeneous media.

system has surfaces separating different media, while it would be a continuous function of position in case our system happens to be an inhomogeneous medium with a continuously varying refractive index.

We start with Fermat's principle, which asserts that the optical path length between any two points P and Q of an actual ray, i.e.,

$$\int_P^Q n(x, y, z) \, ds \qquad (9.1)$$

(where ds represents an infinitesimal change in arc length), is an extremum, i.e.,

$$\delta \int_P^Q n(x, y, z) \, ds = 0 \qquad (9.2)$$

This principle resembles Hamilton's principle of least action in classical mechanics which determines the path along which a system would proceed in time (see, for example, Goldstein, 1950). The similarity between the two may be made more explicit by writing

$$\begin{aligned}
ds &= [(dx)^2 + (dy)^2 + (dz)^2]^{1/2} \\
&= dz\left[1 + \left(\frac{dx}{dz}\right)^2 + \left(\frac{dy}{dz}\right)^2\right]^{1/2}
\end{aligned} \qquad (9.3)$$

Thus we get

$$\delta \int_P^Q n(x, y, z)(1 + \dot{x}^2 + \dot{y}^2)^{1/2} \, dz = 0 \qquad (9.4)$$

where dots represent differentiation with respect to z. The parameter t appearing in Hamilton's principle has been replaced here by the position coordinate z. Here we assume z to represent the approximate direction of propagation. One may, following the definition of Lagrangian in classical mechanics, define an optical Lagrangian L as

$$L = n(x, y, z)(1 + \dot{x}^2 + \dot{y}^2)^{1/2} \qquad (9.5)$$

Applying the same method as in classical mechanics to obtain the Lagrangian equations of motion, one gets the optical Lagrangian equations of motion as

$$\frac{d}{dz}\left(\frac{\partial L}{\partial \dot{x}}\right) = \frac{\partial L}{\partial x} \qquad (9.6)$$

$$\frac{d}{dz}\left(\frac{\partial L}{\partial \dot{y}}\right) = \frac{\partial L}{\partial y} \qquad (9.7)$$

From here we go on to the Hamiltonian formulation; we first define the "conjugate momenta" for x and y as

$$p = \frac{\partial L}{\partial \dot{x}} \tag{9.8}$$

$$q = \frac{\partial L}{\partial \dot{y}} \tag{9.9}$$

The physical significance of p and q can be seen by substituting the value of L from Eq. (9.5) in Eq. (9.8),

$$p = \frac{n\dot{x}}{(1 + \dot{x}^2 + \dot{y}^2)^{1/2}} = n\frac{dx}{ds} \tag{9.10}$$

where dx/ds represents the direction cosine of the ray along the x axis at the point (x, y, z). The direction cosine multiplied by the refractive index at that point, $n(x, y, z)$, would be termed the optical direction cosine. Thus p and q represent the optical direction cosines of the ray along the x and y axes, respectively.

The Lagrangian L is a function of $(x, y, \dot{x}, \dot{y}, z)$. We now switch over to the Hamiltonian formulation, where the Hamiltonian H, assumed to be a function of (x, y, p, q, z), is defined through the relation

$$H(x, y, p, q, z) = p\dot{x} + q\dot{y} - L(x, y, \dot{x}, \dot{y}, z) \tag{9.11}$$

Thus

$$dH = \dot{x}\, dp + \dot{y}\, dq + \left(p - \frac{\partial L}{\partial \dot{x}}\right) d\dot{x} + \left(q - \frac{\partial L}{\partial \dot{y}}\right) d\dot{y} - \frac{\partial L}{\partial x}\, dx - \frac{\partial L}{\partial y}\, dy - \frac{\partial L}{\partial z}\, dz$$

$$= \dot{x}\, dp + \dot{y}\, dq - \dot{p}\, dx - \dot{q}\, dy - \frac{\partial L}{\partial z}\, dz \tag{9.12}$$

where we have used Eqs. (9.6)–(9.9). Further dH is also given by

$$dH = \frac{\partial H}{\partial p}\, dp + \frac{\partial H}{\partial q}\, dq + \frac{\partial H}{\partial x}\, dx + \frac{\partial H}{\partial y}\, dy + \frac{\partial H}{\partial z}\, dz \tag{9.13}$$

Comparing Eqs. (9.12) and (9.13) we obtain the Hamilton's equations:

$$\dot{x} = \frac{dx}{dz} = \frac{\partial H}{\partial p} \tag{9.14}$$

$$\dot{y} = \frac{dy}{dz} = \frac{\partial H}{\partial q} \tag{9.15}$$

$$\dot{p} = \frac{dp}{dz} = -\frac{\partial H}{\partial x} \tag{9.16}$$

$$\dot{q} = \frac{dq}{dz} = -\frac{\partial H}{\partial y} \tag{9.17}$$

and

$$\frac{\partial H}{\partial z} = -\frac{\partial L}{\partial z} \tag{9.18}$$

The above equations form the fundamental equations of the Hamiltonian formulation.

The Hamiltonian H can be put in a more explicit form by noting that

$$p\dot{x} + q\dot{y} - L = \frac{n\dot{x}^2}{(1+\dot{x}^2+\dot{y}^2)^{1/2}} + \frac{n\dot{y}^2}{(1+\dot{x}^2+\dot{y}^2)^{1/2}} - n(1+\dot{x}^2+\dot{y}^2)^{1/2} \tag{9.19}$$

where p has been replaced by its equivalent value given in Eq. (9.10) and q by an exactly similar equation. Hence,

$$H = -\frac{n}{(1+\dot{x}^2+\dot{y}^2)^{1/2}} \tag{9.20}$$

We also note that

$$n^2 - p^2 - q^2 = n^2\left(1 - \frac{\dot{x}^2}{1+\dot{x}^2+\dot{y}^2} - \frac{\dot{y}^2}{1+\dot{x}^2+\dot{y}^2}\right)$$
$$= \frac{n^2}{1+\dot{x}^2+\dot{y}^2} \tag{9.21}$$

Hence, from Eqs. (9.20) and (9.21) we get

$$H = -(n^2 - p^2 - q^2)^{1/2} \tag{9.22}$$

Thus, given an optical system (i.e., given the function n in terms of x, y, z) all one has to do is to calculate H from Eq. (9.22) and use Hamilton's equations to solve for the dependence of x, y, p, and q, on z. Once this is done, the image of any object can be reconstructed. This process is not as simple as has been stated above. For a general system, since H is an irrational function, simple meaningful solutions for x, y as functions of z might not exist, except possibly for some simple systems. So one is led to considering approximate solutions to Hamilton's equations and to better the approximation through consideration of larger and larger numbers of terms.

Since most of the optical systems encountered in practice are cylindrically symmetric we will be dealing here only with such systems; the axis of symmetry, say z, being taken as the direction of propagation. The transverse directions are given by x and y. Clearly the x, y dependence of refractive index $n(x, y, z)$ of such systems should be in the form of $x^2 + y^2$.

Let us put

$$u = x^2 + y^2 \qquad (9.23)$$

$$v = p^2 + q^2 \qquad (9.24)$$

Thus, for cylindrically symmetric systems

$$H = -[n^2(u, z) - v]^{1/2} \qquad (9.25)$$

Hamilton's equations may also be written in terms of u and v by noting that

$$\frac{\partial}{\partial p} = 2p \frac{\partial}{\partial v}$$
$$\frac{\partial}{\partial q} = 2q \frac{\partial}{\partial v} \qquad (9.26)$$

and

$$\frac{\partial}{\partial x} = 2x \frac{\partial}{\partial u}$$
$$\frac{\partial}{\partial y} = 2y \frac{\partial}{\partial u} \qquad (9.27)$$

Hence we get from Eqs. (9.14)–(9.18)

$$\frac{dx}{dz} = 2p \frac{\partial H}{\partial v} \qquad (9.28a)$$

$$\frac{dy}{dz} = 2q \frac{\partial H}{\partial v} \qquad (9.28b)$$

$$\frac{dp}{dz} = -2x \frac{\partial H}{\partial u} \qquad (9.28c)$$

$$\frac{dq}{dz} = -2y \frac{\partial H}{\partial u} \qquad (9.28d)$$

Since our system is completely symmetric in x and y, and p and q, it is enough if we consider only one set of equations say for x and p, and the equations for y and q would follow from analogy.

As mentioned earlier, since it is not possible to solve the equations exactly, we will proceed to derive approximate solutions. The lowest order approximation would be called the paraxial approximation. The optics obtained under such an approximation is termed paraxial optics or Gaussian optics. This approximation is determined by the condition that the ray, as it traverses the system, is always infinitesimally close to the symmetry axis (which is the z axis) and also that the slope of the ray with respect to the z axis is infinitesimal. Under such an approximation the images formed would be perfect, i.e., the image of a point object would be a point. The deviation from this approximation would determine the aberrations present in the medium; different order terms giving rise to different orders of aberration.

To be able to distinguish between paraxial and nonparaxial rays we will from now on employ capital letters for nonparaxial rays and small letters for paraxial rays. Hence the variables for nonparaxial rays would be given by X, Y, P, and Q, while these would be x, y, p, and q for paraxial rays. The capital letters would reduce to the small letters under the paraxial approximation.

For nonparaxial rays the Hamiltonian H and the Hamilton's equations would be

$$H(U, V, z) = -[n^2(U, z) - V]^{1/2} \qquad (9.29)$$

$$\frac{dX}{dz} = 2P\frac{\partial H}{\partial V} \qquad (9.30a)$$

$$\frac{dP}{dz} = -2X\frac{\partial H}{\partial U} \qquad (9.30b)$$

where $U = X^2 + Y^2$ and $V = P^2 + Q^2$.

To be able to specify one ray out of all the possible rays that can pass through the system, one has either to specify the value of (x, y, p, q) at a plane, say $z = z_0$, or one has to specify the values of x and y at two parallel planes: say at $z = z_0$ and $z = \zeta$. The plane $z = \zeta$ is just another reference plane. Here we choose the second set of boundary conditions. Let (x_0, y_0) be the coordinates of the ray on the plane $z = z_0$ and (ξ, η) the coordinates on the plane $z = \zeta$, i.e.,

$$X(z_0) = x_0 \qquad Y(z_0) = y_0$$
$$X(\zeta) = \xi \qquad Y(\zeta) = \eta \qquad (9.31)$$

X, Y, P, and Q will, in general, be functions of x_0, y_0, ξ, η, and of course z. Paraxial rays would be represented by infinitesimal values of x_0, y_0, ξ, and η. Hence we expand X and P in ascending powers of x_0, y_0, ξ, and η to get

$$X = X_1 + X_2 + X_3 + \cdots \qquad (9.32a)$$

$$P = P_1 + P_2 + P_3 + \cdots \qquad (9.32b)$$

where the subscripts represent the order of the term; X_1 being linear in x_0, y_0, ξ, and η, X_2 quadratic in the same variables and so on.

Since our system has been assumed to be cylindrically symmetric, a change of variables from (x_0, y_0, ξ, η) to $(-x_0, -y_0, -\xi, -\eta)$ should only change the sign in X, the magnitude remaining constant. But the even-order terms remain unchanged under such a transformation. Thus the symmetry of the system prohibits the presence of even-order terms in the expansion of X and P in ascending powers of (x_0, y_0, ξ, η). Hence we get

$$X = X_1 + X_3 + \cdots \tag{9.33a}$$

$$P = P_1 + P_3 + \cdots \tag{9.33b}$$

X_1 and P_1 will of course, represent paraxial values of X and P. $X_3 + X_5 + \cdots$ will represent the aberration in the image; X_3 being the third-order aberration since it is of degree three, X_5 the fifth-order aberration and so on. Our aim in this chapter is to calculate X_3 and the corresponding third-order aberration along the y direction, Y_3.

We can also expand $H(U, V, z)$ given by Eq. (9.29) in ascending powers of U and V to get*

$$H = H_0 + H_1 U + H_2 V + \tfrac{1}{2}(H_{11} U^2 + 2H_{12} UV + H_{22} V^2) + \cdots \tag{9.34}$$

where H_i and H_{ij} are all functions of z.

We now substitute the expansions given by Eqs. (9.33a), (9.33b), and (9.34) in Eqs. (9.30a) to get

$$(\dot{X}_1 + \dot{X}_3 + \cdots) = 2(P_1 + P_3 + \cdots)(H_2 + H_{12} U + H_{22} V + \cdots) \tag{9.35}$$

where, as before, dots represent derivatives with respect to z. Equating terms of equal degree we get

$$\dot{X}_1 = 2H_2 P_1 \tag{9.36}$$

$$\dot{X}_3 = 2H_2 P_3 + 2(H_{12} U_1 + H_{22} V_1)P_1 \tag{9.37}$$

where, in Eq. (9.37), U and V have been replaced by $U_1(=X_1^2 + Y_1^2)$ and $V_1(=P_1^2 + Q_1^2)$, respectively. Similarly, using Eq. (9.30b), we get

$$\dot{P}_1 = -2H_1 X_1 \tag{9.38}$$

$$\dot{P}_3 = -2H_1 X_3 - 2(H_{11} U_1 + H_{12} V_1)X_1 \tag{9.39}$$

*$V = p^2 + q^2 = n^2(\cos^2 \alpha + \cos^2 \beta)$, where α and β are the angles that the ray makes with the x and y axes. If γ is the angle made with z axis, then it is known that $\cos^2 \gamma = 1 - (\cos^2 \alpha + \cos^2 \beta)$ and hence $V = n^2(1 - \cos^2 \gamma)$. For paraxial rays, γ has to be infinitesimal and hence V is an infinitesimal quantity. Thus one is allowed to expand in ascending powers of V.

Equations (9.36) and (9.38) would determine the paraxial system of rays and Eqs. (9.37) and (9.39) determine the third-order aberration present in the system. We have exactly analogous equations for Y_1, Y_3, Q_1, and Q_3, that may be written down as

$$\dot{Y}_1 = 2H_2Q_1 \tag{9.40}$$

$$\dot{Y}_3 = 2H_2Q_3 + 2(H_{12}U_1 + H_{22}V_1)Q_1 \tag{9.41}$$

$$\dot{Q}_1 = -2H_1Y_1 \tag{9.42}$$

$$\dot{Q}_3 = -2H_1Y_3 - 2(H_{11}U_1 + H_{12}V_1)Y_1 \tag{9.43}$$

Thus given an optical system one can calculate the quantities H_1, H_2, H_{11}, H_{12}, and H_{22}; then using the first-order equations, the z dependence of X_1 and P_1 can be obtained; finally, using these and Eqs. (9.37) and (9.39) X_3 and P_3 can be calculated.

Instead of leaving the analysis at this stage we would derive explicit expressions for the third-order aberration in terms of the expressions for two paraxial rays obeying certain specific boundary conditions. The two paraxial rays would be specified as solutions of Eqs. (9.36) and (9.38) under the following boundary conditions:

(i) $x(z_0) = 0$, $y(z_0) = 0$, $x(\zeta) = 1$, $y(\zeta) = 0$; where as before, $z = z_0$ is the object plane and $z = \zeta$ any other reference plane. The solution under this boundary condition would be represented by $x(z) \equiv h(z)$, $p(z) \equiv \vartheta(z)$ so that $h(z_0) = 0$, $h(\zeta) = 1$. (Since X_1 and P_1 represent paraxial values, we have replaced them by small letters.)

(ii) $x(z_0) = 1$, $y(z_0) = 0$, $x(\zeta) = 0$, $y(\zeta) = 0$. The solution under this boundary condition would be represented by $x(z) \equiv H(z)$ and $p(z) \equiv \theta(z)$ so that $H(z_0) = 1$, $H(\zeta) = 0$.

These two rays have been shown in Fig. 9.1. The first one is termed axial ray and the second field ray. Now, any ray can be expressed in terms of these two rays, since these two form two linearly independent solutions to the problem.* For example, if one wants to specify a ray which satisfies the conditions $x(z_0) = x_0$, $y(z_0) = y_0$, $x(\zeta) = \xi$, $y(\zeta) = \eta$, then one can at once write

$$x(z) = x_0H(z) + \xi h(z) \tag{9.44}$$

$$y(z) = y_0H(z) + \eta h(z) \tag{9.45}$$

*This is analogous to obtaining any general solution of a second-order differential equation in terms of two linearly independent solutions. Note that here also we have a second-order differential equation obtained from a system of two coupled first-order differential equations.

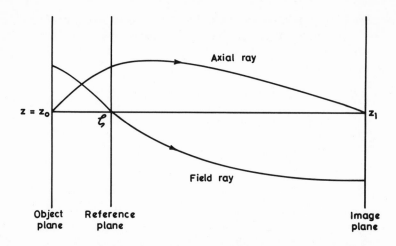

Fig. 9.1. The axial and the field rays are shown. The plane $z = z_1$ is the paraxial image plane. Note that the axial ray intersects the paraxial image plane on the axis. The figure is exaggerated since the two rays shown are supposed to be paraxial rays.

This can be verified by observing that $H(z_0) = 1$, $H(\zeta) = 0$ and $h(z_0) = 0$, $h(\zeta) = 1$. Also since [cf. Eqs. (9.36) and (9.38)]:

$$p = \frac{1}{2H_2}\dot{x} \qquad (9.46a)$$

$$x = -\frac{1}{2H_1}\dot{p} \qquad (9.46b)$$

we have

$$\vartheta = \frac{1}{2H_2}\dot{h} \qquad (9.47a)$$

$$h = -\frac{1}{2H_1}\dot{\vartheta} \qquad (9.47b)$$

and

$$\theta = \frac{1}{2H_2}\dot{H} \qquad (9.48a)$$

$$H = -\frac{1}{2H_1}\dot{\theta} \qquad (9.48b)$$

Hence, differentiating Eqs. (9.44) and (9.45) and dividing both sides by $2H_2$ one obtains

$$p(z) = x_0\theta(z) + \xi\vartheta(z) \qquad (9.49)$$

$$q(z) = y_0\theta(z) + \eta\vartheta(z) \qquad (9.50)$$

Thus the coordinates of any paraxial ray can be expressed in terms of two specific paraxial rays.

We now proceed to calculate the aberration, X_3 and Y_3, in terms of the functions $h(z)$, $H(z)$, $\vartheta(z)$, and $\theta(z)$. The aberration will have to be calculated in a plane at which the image would have formed in the absence of aberrations. This plane is called the paraxial image plane which is determined in the following manner. We know that any ray starting from an axial object point and proceeding along the axis of symmetry will proceed without any deviation. Also, $h(z)$ represents the x coordinate of a ray as a function of z when the object point is axial (because $h(z_0) = 0$; $z = z_0$ being the object plane). Thus the paraxial image point of this object point would be determined by a value $z = z_1$ such that $h(z_1) = 0$. Thus the paraxial image plane is determined from the condition that $h(z_1) = 0$. We have to calculate the aberrations in this plane.

We now have to find the aberrations of a general nonparaxial ray with coordinates (x_0, y_0) in the plane $z = z_0$ and (ξ, η) in the plane $z = \zeta$. The paraxial approximation of such a ray can be described by

$$x(z) = x_0H(z) + \xi h(z) \qquad (9.51a)$$

$$p(z) = x_0\theta(z) + \xi\vartheta(z) \qquad (9.51b)$$

$$y(z) = y_0H(z) + \eta h(z) \qquad (9.51c)$$

$$q(z) = y_0\theta(z) + \eta\vartheta(z) \qquad (9.51d)$$

as has already been shown [Eqs. (9.44), (9.45), (9.49), and (9.50)]. The nonparaxial ray satisfies the conditions,

$$X(z_0) = x_0 \qquad Y(z_0) = y_0 \qquad (9.52a)$$

$$X(\zeta) = \xi \qquad Y(\zeta) = \eta \qquad (9.52b)$$

Looking at Eqs. (9.51) and (9.52) one can see that, both paraxially and nonparaxially, the coordinates of the ray in the plane $z = z_0$ are (x_0, y_0) and in the plane $z = \zeta$ are (ξ, η). Thus we see that the boundary conditions for the paraxial equations are such that

$$X_3(z_0) = 0 \qquad X_3(\zeta) = 0 \qquad (9.53)$$

$$Y_3(z_0) = 0 \qquad Y_3(\zeta) = 0 \qquad (9.54)$$

We now substitute the value of $2H_2$ from Eq. (9.47a) in Eq. (9.37) to get

$$\dot{X}_3\vartheta - \dot{h}P_3 = 2(H_{12}U_1 + H_{22}V_1)p\vartheta \qquad (9.55)$$

where P_1 has been replaced by its paraxial value, p. Similarly substituting for $2H_1$ from Eq. (9.47b) in Eq. (9.39) we obtain

$$\dot{\vartheta}X_3 - h\dot{P}_3 = 2(H_{11}U_1 + H_{12}V_1)xh \qquad (9.56)$$

where X_1 has been replaced by its paraxial value, x. Adding Eqs. (9.55) and (9.56) we get

$$\frac{d}{dz}(X_3\vartheta - hP_3) = 2(H_{12}U_1 + H_{22}V_1)p\vartheta + 2(H_{11}U_1 + H_{12}V_1)xh \qquad (9.57)$$

Integrating from the object plane $z = z_0$ to the image plane $z = z_1$ we get

$$(X_3\vartheta - hP_3)\big|_{z_0}^{z_1} = 2\int_{z_0}^{z_1} [(H_{12}U_1 + H_{22}V_1)p\vartheta + (H_{11}U_1 + H_{12}V_1)xh]\,dz \qquad (9.58)$$

We already know that $h(z_0) = 0$, $h(z_1) = 0$, and $X_3(z_0) = 0$ and $X_3(z_1)$ represents the aberration in the image. Hence

$$X_3(z_1) = \frac{2}{\vartheta(z_1)}\int_{z_0}^{z_1} [(H_{12}U_1 + H_{22}V_1)p\vartheta + (H_{11}U_1 + H_{12}V_1)xh]\,dz \qquad (9.59)$$

Similarly for $Y_3(z_1)$, the aberration in the image along the y direction we get, from analogy,

$$Y_3(z_1) = \frac{2}{\vartheta(z_1)}\int_{z_0}^{z_1} [(H_{12}U_1 + H_{22}V_1)q\vartheta + (H_{11}U_1 + H_{12}V_1)yh]\,dz \qquad (9.60)$$

We would now put the aberrations explicitly in terms of the ray coordinates (x_0, y_0, ξ, η). To do this we note that, using Eqs. (9.51a), (9.51c),

$$U_1 = X_1^2 + Y_1^2$$
$$= x^2 + y^2$$
$$= (x_0H + \xi h)^2 + (y_0H + \eta h)^2$$

i.e.,

$$U_1 = H^2 r + Hht + h^2 s \qquad (9.61)$$

where

$$r = x_0^2 + y_0^2 \tag{9.62}$$

$$s = \xi^2 + \eta^2 \tag{9.63}$$

$$t = 2(x_0\xi + y_0\eta) \tag{9.64}$$

Similarly, using Eqs. (9.51b) and (9.51d) we get

$$V_1 = P_1^2 + Q_1^2$$
$$= p^2 + q^2$$
$$= (x_0\theta + \xi\vartheta)^2 + (y_0\theta + \eta\vartheta)^2$$

i.e.,

$$V_1 = \theta^2 r + \theta\vartheta t + \vartheta^2 s \tag{9.65}$$

Substituting the values of x, p, y, and q from Eqs. (9.51) and the values of U_1 and V_1 from Eqs. (9.61) and (9.65) in Eqs. (9.59) and (9.60) we get

$$\Delta x_1 \equiv X_3(z_1) = [Er + \tfrac{1}{3}Bs + \tfrac{1}{2}(C-D)t]x_0 + (Dr + As + \tfrac{1}{3}Bt)\xi \tag{9.66}$$

$$\Delta y_1 \equiv Y_3(z_1) = [Er + \tfrac{1}{3}Bs + \tfrac{1}{2}(C-D)t]y_0 + (Dr + As + \tfrac{1}{3}Bt)\eta \tag{9.67}$$

where the aberration coefficients A, B, C, D, and E are defined by

$$A = \frac{2}{\vartheta(z_1)} \int_{z_0}^{z_1} (H_{11}h^4 + 2H_{12}h^2\vartheta^2 + H_{22}\vartheta^4)\, dz \tag{9.68}$$

$$B = \frac{6}{\vartheta(z_1)} \int_{z_0}^{z_1} [H_{11}h^3H + H_{12}h\vartheta(H\vartheta + h\theta) + H_{22}\theta\vartheta^3]\, dz \tag{9.69}$$

$$C = \frac{6}{\vartheta(z_1)} \int_{z_0}^{z_1} (H_{11}h^2H^2 + 2H_{12}Hh\theta\vartheta + H_{22}\theta^2\vartheta^2)\, dz + \frac{2}{\vartheta(z_1)}\Gamma^2 \int_{z_0}^{z_1} H_{12}\, dz \tag{9.70}$$

$$D = \frac{2}{\vartheta(z_1)} \int_{z_0}^{z_1} [H_{11}h^2H^2 + H_{12}(H^2\vartheta^2 + h^2\theta^2) + H_{22}\theta^2\vartheta^2]\, dz \tag{9.71}$$

$$E = \frac{2}{\vartheta(z_1)} \int_{z_0}^{z_1} [H_{11}hH^3 + H_{12}H\theta(H\vartheta + h\theta) + H_{22}\theta^3\vartheta]\, dz \tag{9.72}$$

where $\Gamma = H\vartheta - h\theta$ is called the paraxial invariant. In deriving the equation for the coefficient C we have used the relation

$$\frac{d\Gamma}{dz} \equiv \frac{d}{dz}(H\vartheta - h\theta) = 0 \qquad (9.73)$$

This follows easily from Eqs. (9.47a)–(9.48b). It is called the paraxial invariant of the system for the simple reason, as can be seen from Eq. (9.73), that it is invariant with respect to a change of z.

The aberration has been put in the form given by Eqs. (9.66) and (9.67) because, as can be derived from the rigorous formulation (involving characteristic functions), that the generic form of the aberration for any cylindrically symmetric optical system has to be in the form of Eqs. (9.66) and (9.67), where A, B, C, D, and E represent the five Seidel aberrations, namely spherical aberration, coma, astigmatism, curvature of field, and distortion, respectively. These are also the only five types of aberrations required to completely specify the third-order aberration in the image formed by a cylindrically symmetric system. The details of the derivation of this generic form may be found in Luneburg (1964) or Buchdahl (1970). Here we summarize the quantities one needs to know to calculate explicitly the aberration coefficients of a system.

(a) h, ϑ, H, θ. These are obtained as solutions of the paraxial equations under the boundary conditions already given above. One needs to know the values of H_1 and H_2 to be able to solve the paraxial equations. These can be obtained if we note that

$$H_1 = \frac{\partial H}{\partial U}\bigg|_{U=0, V=0} \qquad (9.74)$$

and

$$H_2 = \frac{\partial H}{\partial V}\bigg|_{U=0, V=0} \qquad (9.75)$$

Also

$$H = -[n^2(U, z) - V]^{1/2} \qquad (9.76)$$

Hence

$$H_1 = -\frac{\partial n}{\partial U}\bigg|_{U=0} \qquad (9.77)$$

and

$$H_2 = \frac{1}{2n(o, z)} \qquad (9.78)$$

where $n(o, z)$ is the axial index distribution.

(b) H_{11}, H_{12}, and H_{22}. Since these are coefficients of the Taylor series expansion of H in terms of u and v we have

$$H_{11} = \frac{\partial^2 H}{\partial U^2}\bigg|_{U=0, V=0} \tag{9.79}$$

$$H_{12} = \frac{\partial^2 H}{\partial U \partial V}\bigg|_{U=0, V=0} \tag{9.80}$$

$$H_{22} = \frac{\partial^2 H}{\partial V^2}\bigg|_{U=0, V=0} \tag{9.81}$$

Again using the expression for H [cf. Eq. (9.76)] we get

$$H_{11} = -\frac{\partial^2 n}{\partial U^2}\bigg|_{U=0} \tag{9.82}$$

$$H_{12} = -\frac{1}{2n^2(o, z)}\left(\frac{\partial n}{\partial U}\right)\bigg|_{U=0} \tag{9.83}$$

$$H_{22} = \frac{1}{4n^3(o, z)} \tag{9.84}$$

Thus given an index distribution these quantities can easily be calculated. Explicit calculation of aberrations for a specific system is given in Sec. 9.3.

9.2. Ray Tracing Method

In the previous section, we had derived explicit expressions for third-order aberrations of systems possessing cylindrical symmetry. For systems not possessing such symmetry properties, the generalization of the type of analysis we have presented becomes very tedious. In such cases one would like to resort to the ray tracing technique. This technique involves step by step tracing of rays through the system from the object point to the image plane and determining the deviation from the value predicted by paraxial optics, which gives the aberration. This technique has certain advantages, viz., one can calculate the total aberration present in the ray (i.e., the sum of third-order, fifth-order, etc.) as compared to the analytical technique derived in the last section which gives just the third-order aberrations. At the same time, when the numerical method is used one does not know the distribution of the total aberration between the different elements involved in the optical system; whereas the analytical method gives the element by element contribution to the aberration and hence is useful if some counteractive measures have to be taken to decrease the aberration. Although the numerical ray tracing technique is very time consuming it can be applied to any general variation of the refractive index.

The ray tracing technique involves step by step integration of the ray equation. The ray equation

$$\frac{d}{ds}\left(n\frac{d\mathbf{r}}{ds}\right) = \nabla n \tag{9.85}$$

can be obtained from the Lagrangian formulation given in Sec. 9.1 (see Appendix D).

Given a refractive index distribution $n(x, y, z)$ one has, in general, to solve the three scalar equations given by Eq. (9.85) to get the paths of all rays and hence the aberrations. But the solution of Eq. (9.85) cannot, in general, be obtained for any refractive index function $n(x, y, z)$. One can devise a program for the computer to get the path of all rays, in general. Montagnino (1968) has suggested the following method.

The tracing of a ray through a system means that given a ray at any point specified by its coordinates (spatial coordinates and direction cosines) one should find the coordinates of the ray at a neighboring point such that it satisfies Eq. (9.85). Let us expand $\mathbf{r}$ in a Taylor series in terms of the arc length s about the starting point P_0. Then

$$\mathbf{r} = \mathbf{r}(s_0) + \frac{d\mathbf{r}(s_0)}{ds}\Delta s + \frac{1}{2}\frac{d^2\mathbf{r}(s_0)}{ds^2}(\Delta s)^2 + \cdots \tag{9.86}$$

where $[d\mathbf{r}(s_0)/ds] = (d\mathbf{r}/ds)|_{s=s_0}$, etc. We can similarly expand the derivative $d\mathbf{r}/ds$ in terms of s to get

$$\frac{d\mathbf{r}}{ds} = \frac{d\mathbf{r}(s_0)}{ds} + \frac{d^2\mathbf{r}(s_0)}{ds^2}\Delta s + \cdots \tag{9.87}$$

$d\mathbf{r}/ds$ would represent the unit vector tangential to the ray, denoted by $\hat{\mathbf{t}}$, and $d^2\mathbf{r}/ds^2$ would represent the curvature vector of the ray, represented by $\mathbf{K}$. ($\mathbf{K}$ will represent the rate of change of slope with s.) It can be seen that $\mathbf{K}$ and $\hat{\mathbf{t}}$ are normal to each other so that

$$\mathbf{K} \cdot \hat{\mathbf{t}} = 0 \tag{9.88}$$

Using these, Eq. (9.87) reduces to

$$\hat{\mathbf{t}}(s) = \hat{\mathbf{t}}(s_0) + \mathbf{K}(s_0)\Delta s + \cdots \tag{9.89}$$

Equation (9.85) gives

$$\frac{dn}{ds}\frac{d\mathbf{r}}{ds} + n\frac{d^2\mathbf{r}}{ds^2} = \nabla n$$

or

$$\hat{\mathbf{t}}\frac{dn}{ds}+n\mathbf{K}=\nabla n \tag{9.90}$$

But

$$\frac{dn}{ds}=\hat{\mathbf{t}}\cdot\nabla n \tag{9.91}$$

Hence, using Eqs. (9.90) and (9.91) we get

$$\mathbf{K}=\frac{[\nabla n-\hat{\mathbf{t}}(\hat{\mathbf{t}}\cdot\nabla n)]}{n} \tag{9.92}$$

To trace the ray one proceeds as follows. Given the value of $\mathbf{r}$ and $\hat{\mathbf{t}}(=d\mathbf{r}/ds)$ at a particular point, calculate $\mathbf{K}$ from Eq. (9.92). Then using this value of $\mathbf{K}$, calculate $\mathbf{r}$ and $\hat{\mathbf{t}}$ at a neighboring point by using Eqs. (9.86) and (9.89), for a small value of Δs. By proceeding in this manner one can calculate the values of $\mathbf{r}$ and $\hat{\mathbf{t}}$ for the ray at different points and thus amounting to tracing the ray through the system. Good results can be obtained by taking smaller and smaller values of Δs and carrying out the calculations to many significant places.

The above analysis does not assume any symmetry of $n(x, y, z)$. In cases where such symmetries are present, the problem can be formulated in much simpler ways. In some cases even analytic solutions may be possible. One example where analytic solution is possible is, when the refractive index variation is given by

$$n^2(r)=n_0^2\pm b^2 r^2 \tag{9.93}$$

where $r^2=x^2+y^2$ and z represents the direction of propagation. The negative sign corresponds to the parabolic index medium that finds applications in optical communication systems (see Chapter 5 and Appendix B). Since n is independent of z, the z component of Eq. (9.85) is

$$\frac{d}{ds}\left(n\frac{dz}{ds}\right)=0$$

i.e.,

$$n\frac{dz}{ds}=\text{constant}=n_0\cos\gamma_0=l_0\quad\text{(say)}, \tag{9.94}$$

where n_0 is the refractive index at the starting point of the ray and γ_0 represents the initial angle made by the ray with the z axis. Equation (9.94) is Snell's law for media with continuously varying refractive index.

The x component of Eq. (9.85) is

$$\frac{d}{ds}\left(n\frac{dx}{ds}\right) = \frac{\partial n}{\partial x} \tag{9.95}$$

From Eq. (9.94) it follows that

$$n\frac{d}{ds} = l_0\frac{d}{dz} \tag{9.96}$$

Multiplying both sides of Eq. (9.95) by n, and using Eqs. (9.93) and (9.96) we get

$$\frac{d^2x}{dz^2} = \pm\left(\frac{b}{l_0}\right)^2 x \tag{9.97}$$

Similarly, for the y component we get

$$\frac{d^2y}{dz^2} = \pm\left(\frac{b}{l_0}\right)^2 y \tag{9.98}$$

Equations (9.97) and (9.98) have, as their solutions, linear combinations of hyperbolic sine and cosine functions or a sum of sine and cosine functions. Thus the index distribution given by Eq. (9.93) lends itself to exact solutions. The ray path depends on the initial condition of the ray, i.e., the angle made with the z axis, and hence the system will suffer from aberrations.

It can be shown, analytically, that one can find two different distributions of refractive index n, which are free from any aberration for two different subclasses of rays that can pass through the system.

(a) Meridionally exact distribution, namely,

$$n(r) = n_A \operatorname{sech}(\alpha r)$$
$$= n_A(1 - \tfrac{1}{2}\alpha^2 r^2 + \tfrac{5}{24}\alpha^4 r^4 - \cdots) \tag{9.99}$$

through which meridional rays, i.e., rays lying in a plane containing the optical axis ($r = 0$), pass without aberrations [see, e.g., Kawakami and Nishizawa (1968)].

(b) Helically exact distribution, namely

$$n(r) = \frac{n_A}{(1 + \alpha^2 r^2)^{1/2}}$$
$$= n_A(1 - \tfrac{1}{2}\alpha^2 r^2 + \tfrac{3}{8}\alpha^4 r^4 - \cdots) \tag{9.100}$$

through which helical rays, i.e., a special kind of skew rays which maintain a constant distance from the optical axis, pass without aberrations [see, e.g., Kawakami and Nishizawa (1968)]. Since the two distributions are different, there can be no one distribution which can image objects without aberrations for all sets of rays starting from the object point.

We have as an example of the numerical ray tracing technique the calculations of Rawson *et al.* (1970) who have obtained the variation of aberrations with the initial angle made by the ray with z axis (the optical axis) for different distributions varying from the meridionally exact to the helically exact distributions. Since they have made a numerical computation using a refractive index variation containing terms till the 12th order, their results represent collectively the aberrations till the 11th order.

In the next section we give a comparison of the numerical results so obtained, with the analytical calculations obtained using the formula given in Sec. 9.1.

9.3. Numerical Results

We will now use the theory developed in Sec. 9.1 to calculate third-order aberrations for specific inhomogeneous media. Although we have considered only third-order aberrations, it will be shown that the analytical and numerical calculations match very well, meaning thereby that the contribution from higher order aberrations is very small.* We will also show how zero aberration conditions on special subclasses of rays give us the meridionally exact and helically exact distributions.

We write the refractive index variation of our medium in the form

$$n(U, z) = n_0(1 - \tfrac{1}{2}\alpha^2 U + \tfrac{1}{2}\beta\alpha^4 U^2) \tag{9.101}$$

where α and β are constants of the medium and n_0 is the axial refractive index, assumed to be independent of z and, as before, $u = x^2 + y^2$. The paraxial equations of our system [see Eqs. (9.36) and (9.38)] reduce to

$$\frac{dx}{dz} = \frac{1}{n_0} p \tag{9.102}$$

and

$$\frac{dp}{dz} = -n_0\alpha^2 x$$

The solutions for this set of equations are

$$x = A \sin \alpha z + B \cos \alpha z$$
$$p = n_0\alpha(A \cos \alpha z - B \sin \alpha z) \tag{9.103}$$

*The theory for fifth-order aberrations has been given by Gupta *et al.* (1976).

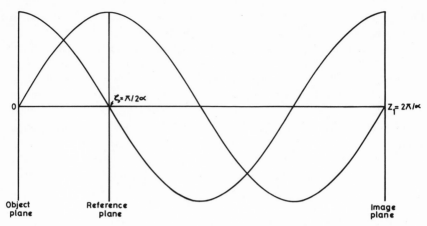

Fig. 9.2. The axial and field rays for a refractive index distribution given by Eq. (9.101).

The axial and the field rays would be given by

$$h(z) = \sin \alpha z \qquad \vartheta(z) = n_0 \alpha \cos \alpha z$$
$$H(z) = \cos \alpha z \qquad \theta(z) = -n_0 \alpha \sin \alpha z$$

(9.104)

where the reference plane $z = \zeta$ has been chosen such that $\sin \alpha \zeta = 1$. These are plotted in Fig. 9.2. The paraxial image planes would be given by

$$\sin \alpha z_1 = 0$$

(9.105)

i.e., $\alpha z_1 = 2m\pi$ $(m = 1, 2, \ldots)$. Choosing $m = 1$, for simplicity, we obtain

$$z_1 = 2\pi/\alpha$$

(9.106)

For the index distribution given by Eq. (9.101), H_{11}, H_{12}, and H_{22} can be obtained

$$H_{11} = -n_0 \beta \alpha^4$$

$$H_{12} = \frac{\alpha^2}{4n_0}$$

(9.107)

$$H_{22} = \frac{1}{4n_0^3}$$

Using the above values of H_{11}, H_{12}, and H_{22} and substituting the expressions for $h(z)$, $\vartheta(z)$, $H(z)$, and $\theta(z)$ from Eq. (9.104) in Eqs. (9.68)–(9.72), one can readily obtain the coefficients A, B, C, D, and E. For example, since $z_0 = 0$ and $z_1 = 2\pi/\alpha$, we have

$$\vartheta(z_1) = n_0 \alpha$$

and

$$A = \frac{2}{n_0 \alpha} \int_0^{2\pi/\alpha} \left(-n_0 \beta \alpha^4 \sin^4 \alpha z + \frac{\alpha^2}{2n_0} n_0^2 \alpha^2 \cos^2 \alpha z \sin^2 \alpha z \right.$$

$$\left. + \frac{1}{4n_0^3} n_0^4 \alpha^4 \cos^4 \alpha z \right) dz$$

$$= z_1 \alpha^3 (\tfrac{5}{16} - \tfrac{3}{4}\beta) \tag{9.108}$$

Similarly

$$B = 0 \tag{9.109}$$

$$C = z_1 \alpha^3 (\tfrac{5}{16} - \tfrac{3}{4}\beta) \tag{9.110}$$

$$D = z_1 \alpha^3 (\tfrac{7}{16} - \tfrac{1}{4}\beta) \tag{9.111}$$

$$E = 0 \tag{9.112}$$

From Eqs. (9.109) and (9.112) it follows that the system is free from coma ($B = 0$) and distortion ($E = 0$). The total aberration [see Eqs. (9.66) and (9.67)] can be obtained as

$$\Delta x_1 = (\tfrac{5}{16} - \tfrac{3}{4}\beta) z_1 \alpha^3 \xi (x_0^2 + \xi^2 + \eta^2) \tag{9.113}$$

$$\Delta y_1 = z_1 \alpha^3 [(\tfrac{7}{16} - \tfrac{1}{4}\beta) x_0^2 + (\tfrac{5}{16} - \tfrac{3}{4}\beta)(\xi^2 + \eta^2)] \eta \tag{9.114}$$

where the object point has been specified by $(x_0, 0)$. This is permissible, without loss of generality, due to the cylindrical symmetry possessed by the system.

We will now determine under what conditions, at least for some special subclass of rays, the aberrations given by Eqs. (9.113) and (9.114) vanish. Let us first consider meridional rays. Since these rays are specified by the condition that the ray remains always in one plane containing the optical axis, as the object point lies in the xz plane, the ray should satisfy the condition $\eta = 0$ in the plane $z = \zeta$. Under this condition, only Δx_1 given by Eq. (9.113) survives; $\Delta y_1 = 0$. It is also clear from the expression for Δx_1 that it reduces to zero when $\beta = \tfrac{5}{12}$. This condition along with Eq. (9.101) can be seen to yield the hyperbolic secant distribution (of course, only the first three terms are obtained since we are only concerned with third-order aberration). Similarly, let us consider a special kind of skew rays, namely helical rays, which are rays which maintain a constant distance from the optical axis. Under the condition that the rays are launched parallel to the yz plane from an off-axis object point, $\xi = 0$ in the plane $z = \zeta$. This makes $\Delta x_1 = 0$, and

$$\Delta y_1 = z_1 \alpha^3 \eta^3 (\tfrac{3}{4} - \beta) \tag{9.115}$$

where we have used the condition that $x_0 = \eta$, i.e., the ray is helical. From

Eq. (9.115) it follows that the ray aberration is identically zero when $\beta = \frac{3}{4}$. This value of β combined with Eq. (9.101) yields the exact distribution for helical rays [Eq. (9.100)]. Thus it is clear from above that certain specific values of β free the meridional and helical rays, separately, of aberrations. It can also be seen that no one value of β can make the aberrations go to zero and hence, there cannot exist any one index distribution which is free from all aberrations for all sets of rays starting from the object point.

To compare the results of the theory developed with those of Rawson *et al.* (1970), we calculate the variation of aberration with the initial angle made by the ray with z axis. The system is assumed to be devoid of any surface between the object and the image plane. The different distributions chosen are: (a) meridionally exact distribution $(\beta = \frac{5}{12})$, and (b) helically exact distribution $(\beta = \frac{3}{4})$. Also drawn are the variation of aberration for (c) parabolic distribution $(\beta = 0)$ and (d) a distribution found in experimentally fabricated rods $(\beta = 0.95)$. We consider two sets of rays (a) meridional and (b) skew. For the case of meridional rays we consider two sets of object points (a) axial object point and (b) an off-axis object point situated at a distance of $x_0 = 0.095 \, z_1$ from the axis, corresponding to the first set of graphs of Rawson *et al.* (1970). For the skew rays we again consider an off-axis object point situated at $x_0 = 0.095 \, z_1$, the rays emanating at different angles in the yz plane. Table 9.1 gives the various formulas used in the calculations; γ_0 represents the initial angle between the ray and the z axis. Figures 9.3, 9.4, and 9.5 show the variation of aberration (in units of z_1) with

Table 9.1. *Expressions for Meridional and Skew Ray Aberrations for Different Distributions*

| Distribution | Meridional rays, $\Delta y_1 = 0$ | | Skew rays, $\Delta x_1 = 0$ |
	Axial object point $(\Delta x_1 =)$	Nonaxial object point $(\Delta x_1 =)$	Nonaxial object point $(\Delta y_1 =)$
Meridional	0	0	$0.1188 \, z_1 \sin \gamma_0$
Helical	$-\frac{1}{4} z_1 \sin^3 \gamma_0$	$-\frac{z_1}{4}(0.3564 + \sin^2 \gamma_0)$ $\times \sin \gamma_0$	$\frac{z_1}{4}(0.3564 - \sin^2 \gamma_0)$ $\times \sin \gamma_0$
Parabolic	$\frac{5 z_1}{16} \sin^3 \gamma_0$	$\frac{5 z_1}{16}(0.3564 + \sin^2 \gamma_0)$ $\times \sin \gamma_0$	$\frac{5 z_1}{16}(0.4991 + \sin^2 \gamma_0)$ $\times \sin \gamma_0$
Experimental	$-\frac{2 z_1}{5} \sin^3 \gamma_0$	$-\frac{2 z_1}{5}(0.3564 + \sin^2 \gamma_0)$ $\times \sin \gamma_0$	$\frac{z_1}{5}(0.3564 - 2 \sin^2 \gamma_0)$ $\times \sin \gamma_0$

γ_0. Also shown, for comparison, are the curves of Rawson *et al.* (1970). The curves show an excellent agreement. The slight disagreement is due to the fact that Rawson *et al.*'s calculations give a cumulative effect of the various orders of aberration. This implies that the contibution from higher order aberrations is small.

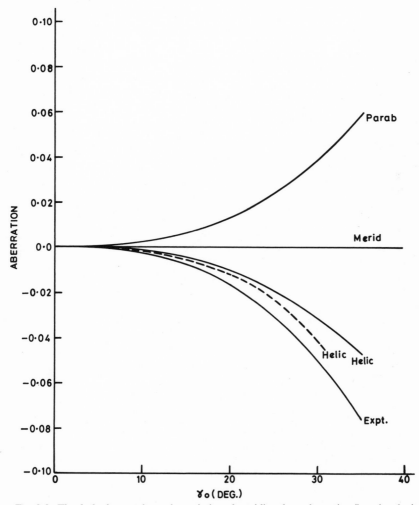

Fig. 9.3. The dashed curve shows the variation of meridional ray aberration (in units of z_1) with γ_0 as obtained by Rawson *et al.* (1970) using numerical techniques; γ_0 represents the initial angle made by the ray with the z axis. The solid curves correspond to the results obtained by using third-order aberration formulas (see Sec. 9.1). The different distributions used are (a) parabolic marked Parab, (b) meridional marked Merid, (c) helical marked Helic, and (d) experimentally prepared rods marked Expt. Rawson *et al.*'s (1970) results for the meridional distribution coincide with that of third-order calculations.

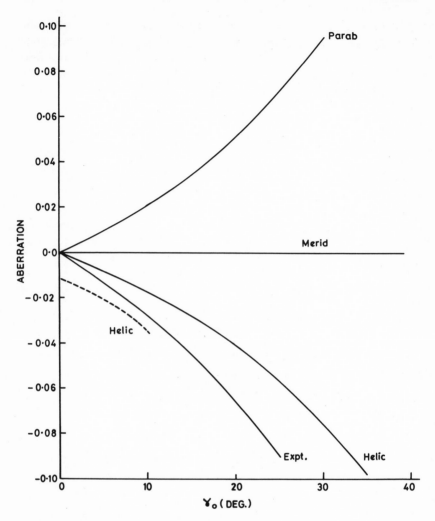

Fig. 9.4. Comparison of meridional ray aberration for a nonaxial object point between Rawson *et al.*'s (1970) numerical calculation (shown as dashed lines) and the results obtained using third-order aberration formulas. The distributions used are the same as those of Fig. 9.3. The meridional ray aberrations agree exactly. The disagreement for helical distribution is due to the presence of higher order aberrations.

In summary, we have developed the Hamiltonian theory of third-order aberrations of media possessing cylindrical symmetry. Explicit expressions have been given for the various Seidel aberrations present. Some analysis of ray tracing technique has also been given with an example of a system where exact analytical results may also be obtained. The theory developed has been used to compare the results predicted for some specific systems with the corresponding ones of Rawson *et al.* (1970) and it is found that third-order theory suffices for practical systems. We have also shown that zero aberration conditions on some specific subclasses of rays yield index distributions which agree with the corresponding ones obtained by other methods.

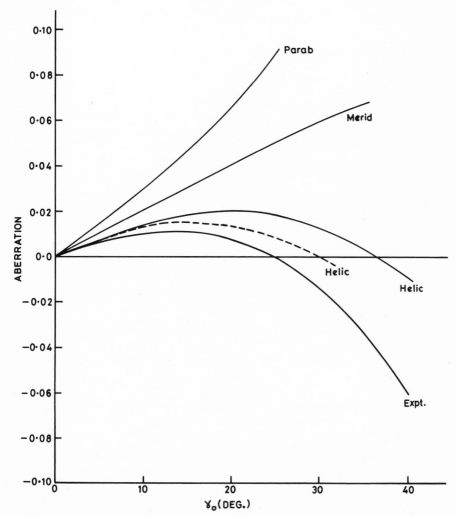

Fig. 9.5. Comparison of skew ray aberrations for a nonaxial object point between Rawson *et al.*'s (1970) numerical calculations, shown dotted and the results obtained using third-order aberration formulas. The distributions used are the same as those of Fig. 9.3. The slight disagreement is due to the presence of higher order aberrations. The meridional ray aberrations agree exactly.

Fabrication of Planar Waveguides

In this appendix we will discuss the various techniques for fabrication of planar waveguides. In planar waveguides the dielectric constant is uniform over a plane and varies only in one direction (perpendicular to the planes).

A.1. Thin-Film Waveguides

A.1.1. Thin-Film Dielectric Waveguides

In the area of integrated optical circuits considerable effort has been directed towards the production of thin film dielectric waveguides. These (see Fig. A.1) essentially consist of a film whose thickness is of the order of 1 μm or less. Such a thin film has to be supported on a substrate. The films used by Goell and Standley (1969) for waveguide fabrication were prepared by sputtering of suitable glasses. The sputtering system used oil-diffusion pumps and 5-in.-diameter electrodes. Oxygen was used as the sputtering gas. Laboratory slides were used as substrates. The films were about 0.3-μm thick, and had a refractive index of about 1.62. The losses were about 1 dB/cm. Tien *et al.* (1969) used sputtered films of ZnO and films of ZnS, evaporated by electron bombardment. The film thickness ranged from 800 to 30,000 Å. They had also used a novel film prism coupler for exciting the

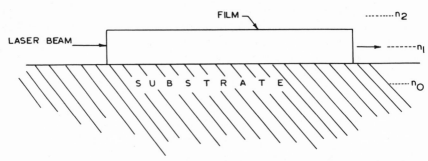

Fig. A.1. The cross section of a thin-film waveguide. The laser beam propagates in the film parallel to the z axis (after Tien, 1971; reprinted by permission).

desired mode of propagating light wave in these films; the theory of the prism film coupler has been discussed by Tien and Ulrich (1970) and Ulrich (1970). Films used in the experiments of Harris *et al.* (1970) were fabricated with both polyester and polyurethane epoxy resins, chosen for their low optical absorption.

Tien *et al.* (1972) have fabricated polymerized organosilicon films on glass substrates in an argon–monomer discharge. The organosilicon films made from vinyl-trimethylsilane (VTMS) and hexamethyl disiloxane (HMDS) monomers have a loss of the order of 0.04 dB/cm; this may be compared to the several hundred times larger loss in some earlier experiments. For the helium–neon laser wavelength (6328 Å), a typical VTMS film was found to have a refractive index of 1.531, which is about 1% larger than that of ordinary glass (1.512). Since for light guiding experiments a film having a refractive index only slightly greater than that of the glass substrate is preferred, VTMS films proved excellent for this use. A typical HMDS film has an index of 1.488, which is smaller than that of ordinary glass but greater than that of Corning 744 Pyrex glass (1.4704). A superposition of a VTMS on an HMDS film thus permits construction of optical circuitry in multilayers. The details of fabrication of such a film and its light propagation characteristics have been given by Tien *et al.* (1972).

A.1.2.   Guided Propagation through Thin Films

As already mentioned, a thin film acts as a dielectric waveguide when the refractive index of the film, n_1, is greater than that of the substrate, n_0, and as well as that of the air space above, n_2 (which is always the case). A typical waveguiding action, as demonstrated by Tien (1971), is shown in Fig. A.2. Here the entire surface of a 7.6×2.5 cm microscope glass slide is coated with a layer of organic film made from vinyltrimethylsilane. Corresponding to a helium–neon laser beam ($\lambda = 6328$ Å), $n_0 = 1.5125$ and $n_1 =$

1.5301. A light beam was fed into the film at the left side of the figure. It propagated through the entire length of the film and then radiated into the free space at the right side of the film. In order to show that the light was truly propagating inside the film, the film was scratched as shown in Fig. A.3. The light beam stopped at the scratched point, which radiated brightly as an antenna (Tien, 1971).

A.1.3.   Asymmetric Slab Waveguides

Asymmetric metal/dielectric/dielectric layer structures have been fabricated and used for guidance of light for quite sometime (see, for example, Otto and Sohler, 1971). In a recent paper Kaminow *et al.* (1974) have reported the fabrication of (a) air/polymer/glass (APG) guides (b) air/polymer/metal (APM) guides, (c) metal/polymer/glass (MPG) guides, and (d) metal/polymer/metal guides, of which the last one happens to have a symmetrical structure (see Fig. 2.2).

In a typical experimental setup (Kaminow *et al.*, 1974), a glass microscope slide $(80 \times 25 \times 1 \text{ mm})$ with refractive index $n_G = 1.513$ $(\lambda = 6330 \text{ Å})$ was cleaned with detergent in an ultrasonic bath and a metal layer with thickness 0.1 μm was evaporated onto it in vacuum. The metal was evaporated through 20×20 mm square masks or stepped masks allowing metal guide lengths of 1, 10, and 20 mm. The polymer layer consisted of

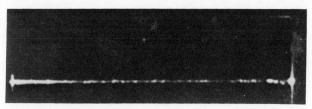

Fig. A.2. Guided propagation of a light beam in a thin film which is coated on a 2.5×7.6 cm microscope glass slide (after Tien, 1971; reprinted by permission).

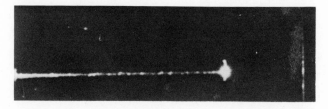

Fig. A.3. To show that the light wave was truly propagating in the film, it was scratched and the light beam was observed to stop immediately (after Tien, 1971; reprinted by permission).

specially prepared low molecular weight photoresist with refractive index $n_P = 1.588$. A solution of photoresist in KPR thinner was applied to the surface of the slide, which was then raised to the vertical position to drain off excess solution. The layer thickness, d, is determined by the polymer concentration in the solution. Values of thickness in the range $1 < d < 10$ μm were achieved. Losses measured in polymer layers 2-μm thick on glass slides were 0.15 dB/cm in the fundamental mode (Kaminow *et al.*, 1974).

A.2.   Optical Waveguides Formed by Electrically Induced Migration of Ions in Glass Plates

Izawa and Nakagome (1972) have reported the fabrication of low-loss ($\leqslant 0.1$ dB/cm) optical waveguides for integrated optical devices by electrically induced migration of ions in glass plates. The process is based on the principle that if an ion with small electronic polarizability replaces an ion with large polarizability the refractive index of the glass can be increased. Thus, the fabrication is carried out in two steps. In the first step, ions with large electronic polarizability are diffused under an electric field and a layer of high refractive index is formed at the surface of the glass substrate. In the second step, ions with small electronic polarizability are diffused, again in the presence of an electric field. Because of the electric field the high refractive index layer (formed in the first step) moves inward and one finally obtains a high refractive index layer sandwiched between two low refractive index layers.

The diffusion of ions in the presence of an electric field is governed by the equation

$$\frac{\partial C}{\partial t} = D \frac{\partial^2 C}{\partial x^2} - E\mu \frac{\partial C}{\partial x} \tag{A.1}$$

where C is the concentration of the diffusing ion, μ the ionic mobility, D the diffusion coefficient, and E the applied electric field. The solution of the above equation, subject to the conditions that $C(x = 0) = 0$ for $x > 0$ and $C(0, t) = C_0$ for $t > 0$, is given by

$$C(x, t) = C_0 \operatorname{erfc}\left[\frac{x - E\mu t}{2(Dt)^{1/2}}\right] \tag{A.2}$$

where the complementary error function $\operatorname{erfc}(\xi)$ is defined by the equation

$$\operatorname{erfc}(\xi) = \frac{2}{\sqrt{\pi}} \int_{\xi}^{\infty} e^{-t^2}\, dt \tag{A.3}$$

and is tabulated at many places.

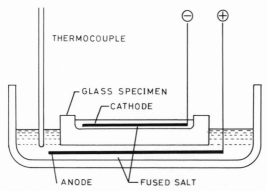

Fig. A.4. Schematic diagram of apparatus for electrically induced migration of ions (after Izawa and
Nakagome, 1972; reprinted by permission).

The schematic diagram of the apparatus used by Izawa and Nakagome
(1972) is shown in Fig. A.4. The substrate used was optically homogeneous
borosilicate glass, which contained potassium oxide and sodium oxide. In
the first step, a mixture of thallium nitrate, sodium nitrate, and potassium
nitrate is used as molten salts, in which the bottom of the substrate is
immersed.* It should be noted that the thallium ion has a large electronic
polarizability; consequently when some parts of the Na^+ and K^+ ions in a
glass are exchanged with Tl^+ ions, a high refractive index layer is formed. In
the second step of the process a molten mixture of only $NaNO_3$ and KNO_3 is
used. The high-index layer moves inward, and a low-index layer is formed at
the surface. The refractive index difference can be adjusted between 0.0005
and 0.1 by changing Tl^+ ion concentration in the molten salts.

The refractive index variation can easily be understood from the
interference pattern (Fig. A.5) observed through an interference microscope. Each fringe corresponds to an optical path difference of $\lambda/2$. Figure
A.6 corresponds to a high refractive index layer of 50-μm thickness formed
about 250 μm beneath the surface.

A.3.   Optical Waveguides of LiNbO₃ and LiTaO₃

In a recent paper Kaminow and Carruthers (1973) have developed an
outdiffusion technique for achieving thin positive-index layers in $LiNbO_3$
and $LiTaO_3$, which are among the best available electro- and acousto-optic
materials.

---

*An electric field of about 10 V/mm is applied between the platinum electrodes. It should be
mentioned that it is not necessary to apply the electric field during the first step; however, by
the application of the electric field the processing time is shortened.

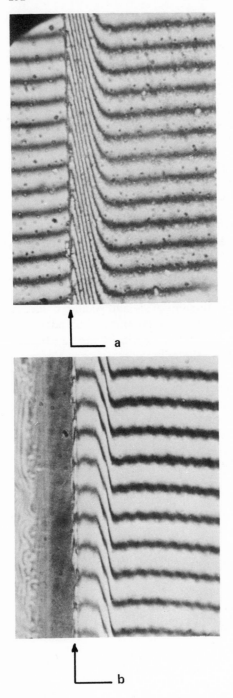

Fig. A.5. Interference patterns observed through an interference microscope. Each fringe corresponds to an optical path difference of a half-wavelength. The sample is 250-μm thick: (a) sample treated in molten salts containing Tl ions at 530 °C for 72 hr with no electric field (the refractive index is increased at the surface); (b) treated at 530°C for 50 min with an electric field of 7 V/cm. A plateau of the high refractive index region is formed at the surface (after Izawa and Nakagome, 1972; reprinted by permission).

surface

Fig. A.6. Interference pattern of the 250-μm thick sample treated in a mixture of TlNO$_3$, KNO$_3$, and NaNO$_3$ for 23 hr with no field, then treated in KNO$_3$ and NaNO$_3$ for 100 min with an electric field of 7 V/cm (after Izawa and Nakagome, 1972; reprinted by permission).

Both LiNbO$_3$ and LiTaO$_3$ can crystallize in a nonstoichiometric form $(Li_2O)_\nu(M_2O_5)_{1-\nu}$ with the ν value lying between 0.48 and 0.50 and M standing for either Nb or Ta. From the available data (Bergman *et al.*, 1968; Ballman *et al.*, 1967) it can be inferred that the ordinary refractive index n_0 is independent of ν, but the extraordinary refractive index varies linearly with ν; in fact

$$\frac{dn_e}{d\nu} = -1.63 \qquad \text{for LiNbO}_3$$

and

$$\frac{dn_e}{d\nu} = -0.85 \qquad \text{for LiTaO}_3$$

(Carruthers *et al.*, 1971; Barns and Carruthers, 1970). Thus, if one is able to reduce the value of ν at the surface, n_e will increase and an optical waveguiding layer would result. In principle, one could either diffuse M_2O_5 (M = Nb or Ta) into the surface or outdiffuse Li_2O to achieve a waveguiding layer. However, since the Li bond is much weaker, it may be easier to outdiffuse Li_2O through the surface by heating Li_2MO_3 in vacuum. This technique has been suggested and successfully exploited by Kaminow and Carruthers (1973).

In a typical experimental arrangement a poled $LiNbO_3$ crystal with dimensions $15 \times 2 \times 5$ mm along the a, b, and c crystal axes, respectively, was placed in a Mettler thermogravimetric microbalance which could monitor the weight of the sample during the process of heating. The sample was kept at $T = 1100°C$ and a pressure of 6×10^{-6} Torr for 21 hr, during which the loss of weight was about 180 μg. The corresponding refractive index variation (normal to the surface) was measured with a Leitz interference microscope and is plotted in Fig. A.7 as curve I-2. Other refractive index profiles obtained under different experimental conditions are also plotted in Fig. A.7. The curves are reasonably well represented by an empirical expression involving the complementary error function (Crank, 1956):

$$\Delta n_e = A \text{ erfc}(x/B) \qquad (A.4)$$

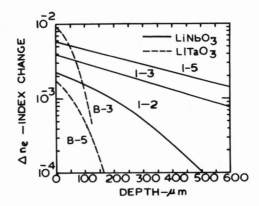

Fig. A.7. Experimental refractive index profiles in $LiNbO_3$ and $LiTaO_3$ for diffusion normal to c (after Kaminow and Carruthers, 1973; reprinted by permission).

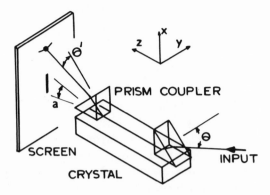

Fig. A.8. Demonstration of optical waveguiding by the prism coupler arrangement (after Kaminow and Carruthers, 1973; reprinted by permission).

where x is the depth below the surface, A is a function of time t and temperature T, and $B = (Dt)^{1/2}$, where D is the diffusion coefficient which is an exponential function of T. For the curve I-2 the values of various constants are given by

$$A = 2.2 \times 10^3$$

$$B = 360 \, \mu$$

$$D = 4.2 \times 10^{-9} \, \text{cm}^2/\text{sec}$$

It can be seen that the curves I-3 and I-5 are approximately straight lines and one may assume the refractive index variation of the form

$$\Delta n_e = A \, \exp(-x/d) \qquad (A.5)$$

A refractive index variation of form (A.4) or (A.5) will guide light near the surface $x = 0$ when A is positive. The waveguiding action on optical beams can be demonstrated by the prism coupler arrangement as shown in Fig. A.8; the orientation of the a, b, and c axes of the crystal has been shown in the diagram. The incident beam is polarized as an extraordinary TE wave and a bright streak appears along the surface when θ is adjusted near an angle θ_0 slightly less than the critical angle. The mode radiates from the end of the guide and the losses reported by Kaminow and Carruthers (1973) are of the order of 2 dB/cm.

A.4. Fabrication of Optical Waveguides by Diffusion

In recent years a considerable amount of experimental work has also been reported on optical waveguides formed by diffusion; this avoids the difficulty of growing suitable thin-film layers. Taylor *et al.* (1972) fabricated optical waveguides by diffusing selenium into single-crystal CdS substrates to produce graded composition CdS_xSe_{1-x} crystals. The CdS substrates, used by Taylor *et al.* (1972) were commercial crystals grown by the Piper–Polich method and polished to a smoothness of $\frac{1}{15}$th of an optical wavelength. The crystals were oriented so that the optic axis was normal to the surface upon which the waveguide was to be fabricated. The diffusion process of Taylor *et al.* (1972) consisted of heating the host crystal in an atmosphere of sulfur and selenium. A quartz tube containing CdS substrate, sulfur, selenium, and CdS powder (added to reduce thermal etching of the substrate during diffusion) was connected to a diffusion pump and sealed at a pressure of 5×10^{-7} Torr. In a typical waveguide fabrication the diffusion time, t, was 40 hr, the temperature T was 600°C, $m_s/v = 1.10\,\text{mg/cm}^3$, and $m_{se}/v = 3.7\,\text{mg/cm}^3$, where m_s/v and m_{se}/v represent the mass of sulfur and selenium per unit ampoule volume. The observed diffusion depth was 2.5 μm.

According to Taylor (1972), the selenium concentration varies with distance below the surface as $\exp(-|x|/d)$, d being the diffusion depth. The departure from the expected Gaussian dependence on x is not unusual, and according to Taylor (1972) it is probably due to the diffusing species being in two forms or occupying two different sites not in local equilibrium. The refractive index has been found to increase linearly with selenium concentration (Lisitsa *et al.*, 1969); the variation of the refractive index is found to be of the form (Conwell, 1973)

$$n(x) = n_0 + \Delta n \, \exp(-|x|/d) \tag{A.6}$$

The corresponding variation of dielectric constant would be given by

$$K(x) = K_0 + \Delta K \, \exp(-|x|/d) \tag{A.7}$$

where $K_0 = n_0^2$; $\Delta K = 2n_0\Delta n$ and we have assumed $\Delta K \ll K_0$. Typically when $d = 2.5\mu$m, $\Delta K = 0.3$ ($K_0 = 6.1$). As shown in Chapter 3, such a system will support eight TE modes for $\lambda = 6330$ Å.

In order to demonstrate the waveguiding action, light from a He-Ne laser was focused onto the polished edge of a diffused crystal with a microscope objective. The diameter of the focused spot was $\sim$3 μm. The beam had spread by diffraction in a direction parallel to the surface of the crystal, but was found to be confined by the waveguide in the direction normal to the surface. Losses were reported to be of the order of 10–15 dB/cm.

A.5.   p–n Junctions as Waveguides

Guided propagation has also been observed in the depletion layer of epitaxially grown and diffusion doped single-crystal semiconductors. Reinhart *et al.*'s (1969) observations of light transmitted along the plane of reverse biased GaP *p–n* junction show the depletion layer to be birefringent and to produce mode confinement of the light. The results have been interpreted in terms of a dielectric waveguide in which a higher optical dielectric constant exists in a layer whose width is fixed and approximately equal to the zero bias junction width. The birefringence arises from the linear electro-optic (Pockels) effect within the junction width, which changes with bias. In the model developed by Reinhart *et al.* (1969) the optical dielectric constants on the *n* and *p* sides are assumed to be slightly different.

Zachos and Ripper (1969) have obtained the resonant modes of GaAs junction lasers from a proposed model and compared their results with experimental observations. The construction of a typical "stripe-geometry laser" is shown in Fig. A.9. An insulating layer of SiO_2 separates the shallow

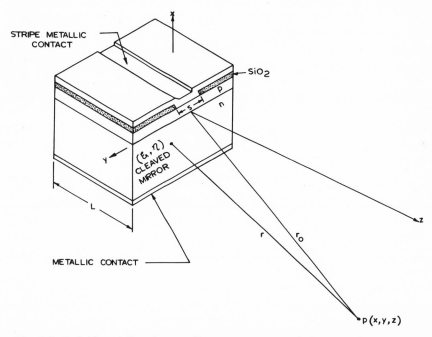

Fig. A.9. Definition of Cartesian coordinate system *x*, *y*, *z* relative to a stripe geometry laser. The *x–y* plane (*z* = 0) coincides with the front laser mirror, which radiates in the half space *z* > 0. *P*(*x*, *y*, *z*) represents a far-field observation point, which is at a distance *r* from a point (ξ, η) on the laser mirror (after Zachos and Ripper, 1969; reprinted by permission).

B.1.   Conventional SELFOC Fibers

Flexible SELFOC fibers are prepared by using the ion-exchange technique. It is well known that any change in the composition or structure of a glass will, in general, produce a refractive index change due to alteration of the electronic structure of the material. Thus one expects that ion exchange will influence the refractive index of the glass. At temperatures above the strain point if a small ion such as Li^+ replaces a larger ion such as Na^+ or K^+, the glass network will relax around the smaller ion to produce a more closely packed structure which usually has a higher refractive index. Conversely, if a large ion replaces a small ion in the glass, the structure can expand to a less packed configuration to give, in most cases, a lower refractive index than that of the original material. Thus, in an alkali ion exchange experiment, one expects to see a refractive index profile near the surface of the glass which follows very closely the concentration profile of the diffusing species. We now proceed to discuss the various aspects relevant to the fabrication of SELFOC fibers.

B.1.1.   Fiber Diameter

Flexibility of a SELFOC fiber is an important characteristic when used as an optical waveguide, and the upper limit of its diameter is determined from its desired flexibility. Experimentally (Kitano *et al.*, 1970) one finds that the minimum radius of curvature R_m without breaking a glass fiber is about 100–150 times its diameter D. Thus, for $R_m = 75$ mm, $D \leqslant 0.5$ mm.

B.1.2.   Change in Refractive Index Induced by Ion-Exchange Methods

The refractive index difference (between the surface and the axis), Δn, induced by the ion-exchange method is estimated by comparing the refractive index of glass containing an oxide of an exchangeable cation with that of the glass containing the same amount of an oxide of other exchangeable cation (Kitano *et al.*, 1970). Tables B.1 and B.2 give the values of the refractive indices of silicate glasses doped with monovalent and divalent ion oxides. Table B.1 shows that ion exchange between any pair of Li^+, Na^+, K^+, Rb^+, and Cs^+ results in a maximum value of $\Delta n = 0.01$–0.03, if all the diffusing ions are substituted for each other. However, ion exchange between Tl^+ and any one of the other cations results in a considerably larger Δn ($= 0.3$ at maximum). On the other hand, although ion exchange between divalent cations seems to result in larger Δn than that between monovalent

Table B.1. Refractive Indices of Silicate Glasses Doped with Monovalent Ion Oxide.[a]

Modifying oxide R_2O	Refractive index	
	SiO_2 70 mole% R_2O 30 mole%	SiO_2 60 mole% CaO 20 mole% R_2O 20 mole%
Li_2O	1.53	1.57
Na_2O	1.50	1.55
K_2O	1.51	1.55
Rb_2O	1.50	1.54
Cs_2O	1.50	1.54
Tl_2O	1.83	1.80

[a] After Kitano *et al.* (1970).

ions, it takes a much longer time in the case of divalent cations. This is due to the fact that diffusion constants of divalent cations are orders of magnitude smaller than those of monovalent cations. Thus the ion exchange between Tl^+ and an alkali ion (like Na^+ or K^+) appears to be the best method for obtaining a large refractive index difference in a short time.

The diffusion of ions can be estimated by solving the diffusion equation. Assuming the length of the glass rod to be infinite one obtains (Kitano *et al.*, 1970):

$$\frac{C(r) - C_0}{C_1 - C_0} = \frac{2}{r_0} \sum_{n=1}^{\infty} \exp(-D\alpha_n^2 t) \frac{J_0(r\alpha_n)}{\alpha_n J_1(r_0\alpha_n)} \tag{B.2}$$

Table B.2. Refractive Indices of Silicate Glasses Doped with Divalent Ion Oxide[a]

Modifying oxide RO	Refractive index
	SiO_2 60 mole% RO 40 mole%
PbO	1.81
BaO	1.68
CdO	1.64
SrO	1.61
CaO	1.59
ZnO	1.58
BeO	1.54
MgO	1.51

[a] After Kitano *et al.* (1970).

where r_0 = radius of the glass rod, C_1 = diffusing ion concentration in the glass at $t = 0$ (diffusing ion concentration in the salt at $t = 0$ is 0), C_0 = diffusion ion concentration on the surface of the rod at $t > 0$, $C(r)$ = distribution of the ion concentration at a distance r from the center, D = diffusion coefficient, and α_n = nth root of the equation $J_0(r_0\alpha) = 0$.

The radial distributions of the concentration as obtained from Eq. (B.2) have been plotted in Fig. B.1 for various values of Dt/r_0^2. Clearly, for large values of t the curve becomes parabolic near $r = 0$ and one should therefore expect a parabolic variation of refractive index near the axis. It should be noted that if the value of the diffusion coefficient increases, the same concentration profile is obtained at a smaller value of time. The value of D for Ca^{++} ions at 645°C is about 3.3×10^{-10} cm^2/sec, whereas that of Na^+ ions in the same mother glass at the same temperature is about 100 times less than that of Ca^{++}; therefore, the diffusion time for Tl^+ is a hundred times that for Na^+ ions for obtaining the same refractive index profile.

B.1.3. Results of Typical Experiments

In the experiment carried out by Kitano *et al.* (1970), a glass fiber (composed of 16% Tl_2O, 12% Na_2O, 24% PbO, and 48% SiO_2 by weight)

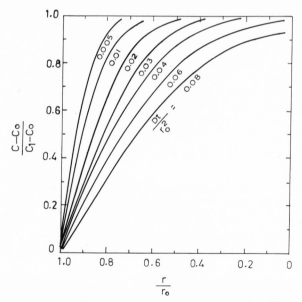

Fig. B.1. Distribution of the ion concentration in glass. D is the diffusion constant, t is time, and r_0 is the radius of a glass fiber (after Kitano *et al.*, 1970; reprinted by permission).

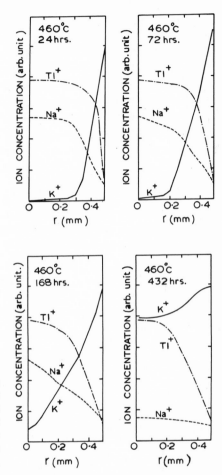

Fig. B.2. The variation of ion concentration distribution in the cross section of the glass rod with time. Ion-exchange temperature is constant, 460°C. The time is 24 hr (a), 72 hr (b), 168 hr (c), and 432 hr (d), and *r* is the distance from the center axis of the rod (after Kitano *et al.*, 1970; reprinted by permission).

was steeped in a KNO_3 bath at 460°C with the aim of making a fiber with $\Delta n = 0.03$. This temperature is between the annealing point and the softening point of the glass. The variation of ion concentration distribution, as a function of the steep time is shown in Fig. B.2. Further soaking makes the distribution of Na^+ very flat and that of K^+ slightly flat, while that of the Tl^+ ion still keeps a parabolic profile. This results from the fact that the diffusion coefficient of Na^+ is slightly larger than that of K^+ ion; both of them are about twice that of Tl^+.

Using the above procedure it was found (Kita and Uchida, 1969; Kitano *et al.*, 1970) that an almost ideal parabolic variation of refractive index was achieved within a 160-μm diameter fiber of 1 m length. The gradient of refractive index variation, a_2, was $0.16 \, \text{mm}^{-2}$.

Figure B.3 shows a typical refractive index profile, obtained by Pearson *et al.* (1969), in a 30 mole% Li_2O, 15 mole% Al_2O_3, 55 mole% SiO_2 glass rod of 1.90 mm diameter, during an ion exchange in a 50 mole% $NaNO_3$, 50 mole% $LiNO_3$ fused salt bath at 470°C for 50 hr. As can be seen from the figure, the refractive index profile is very nearly parabolic. The refractive index at the axis of the rod is approximately 1.539.

The details of the experimental procedure can be found at many places; see, for example, Kitano *et al.* (1970) and Pearson *et al.* (1969).

B.2.   The New SELFOC Fibers

The manufacturing process of the SELFOC fibers as discussed in Sec. B.1 is not only time consuming but it also cannot be used for continuous fabrication of fibers. In a recent paper Koizumi *et al.* (1974) have reported the fabrication of SELFOC fibers by a single continuous process, which makes mass production feasible. Further, the refractive index gradient (i.e., the value of a_2) is so large that even if the fiber is bent randomly the

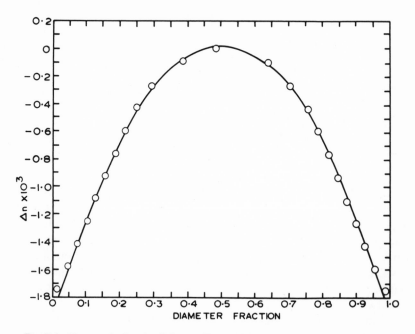

Fig. B.3. Measured refractive index profile of an ion-exchanged rod, normalized to a maximum of zero. The solid line is a parabola fitted to the experimental points by the least-squares method (after Pearson *et al.*, 1969; reprinted by permission).

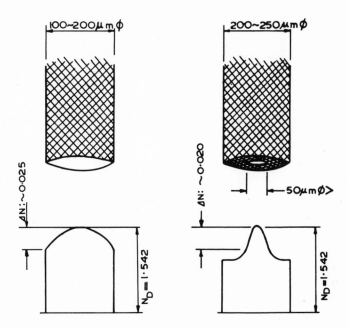

Fig. B.4. Refractive index profiles of the conventional SELFOC fibers and new SELFOC fibers (after Koizumi *et al.*, 1974; reprinted by permission).

propagating laser beam stays near the axis. The core diameters are about 50 μm with outer diameters about 200 μm (thus facilitating greater flexibility) and the value of a_2 ranges from 20 to 300 mm^{-2}. This is to be compared with the SELFOC fibers fabricated by Kitano *et al.* (1970) (using the method described in Sec. B.1), where the outer diameter was about 0.3 mm and the value of $a_2 \approx 0.5$ mm^{-2}. Typical refractive index profiles of the conventional SELFOC fibers and new SELFOC fibers are shown in Fig. B.4. In the fibers fabricated by the new method, group delay resulting from mode conversion is reduced, thus ensuring broad bandwidth.

The new process is shown in Fig. B.5. The inner crucible contains the borosilicate glass with Tl$^+$ to form the core, and the outer crucible contains the borosilicate glass with Na$^+$ to form the cladding. Let the pulling speed of the fiber (of core radius a) be v ($v \approx 20$–40 m/min) then the flowrate V of the core glass flowing downward in the outer crucible is approximately given by

$$\pi R^2 V = \pi a^2 v$$

or

$$V = a^2 v / R^2 \qquad\qquad (B.3)$$

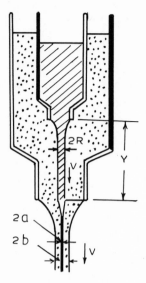

Fig. B.5. Ion-exchange process between core and cladding glasses in a double crucible (after Koizumi *et al.*, 1974; reprinted by permission).

where R is the radius of the core glass flowing downward in the outer crucible. The ion exchange time T is Y/V, where Y is the length along which the core glass contacts the cladding glass. One may define an ion-exchange parameter

$$K = \frac{DT}{R^2} = \frac{DY}{a^2 v} \tag{B.4}$$

Y is the main controllable parameter and in the fibers fabricated by Koizumi *et al.* (1974), the value of K ranged from 0.01 to 0.1.

The details of the transmission characteristics of the new fibers are given in the paper by Koizumi *et al.* (1974); the transmission loss was about 20 dB/km for $\lambda = 0.81$ and 0.85 μm, and a transmission capacity of more than a few G bits/sec/km.

B.3. SELFOC Fibers by Copolymerization

Ohtsuka (1973) has reported the fabrication of a plastic rod having a parabolic refractive index distribution from the copolymerization of diallyl ester (M_1) with vinyl monomer (M_2) whose polymer has a lower refractive index than that of diallyl ester polymer. For M_1 one may use diallyl phthalate, diallyl isophthalate (DAIP), and diethylene glycol bis (allyl carbonate); on the other hand methyl methacrylate (MMA), *n*-butyl methacrylate (BMA), and methyl acrylate (MA) are available for M_2. Ohtsuka (1973) has reported a DAIP–MMA copolymerization technique and has successfully fabricated a light focusing plastic rod having parabolic refractive index variation with the period of the ray path, $L = 50$ mm and $a_2 = 0.0158$ mm^{-2}. The details of the procedure for fabrication can be found in the paper by Ohtsuka (1973).

B.4. Gas Lenses

B.4.1. Experimental Method

A gas lens of laminar flow type consists of a tube that is kept at a constant temperature higher than the temperature of the gas flowing through it. The gas which is close to the walls heats up. The heat penetrates radially into the gas establishing a temperature gradient that causes a corresponding refractive index gradient. The refractive index is higher at the center of the tube and decreases toward the wall. A typical experimental arrangement, used by Aoki and Suzuki (1966), consisted of a brass pipe (of internal diameter 0.85 cm and length 45 cm) with nichrome wire wound around it for warming and an air source attachment (see Fig. B.6). For insulation the brass pipe is covered with mica. The electrical current flowing through the nichrome wire is adjusted by a rheostat and room temperature air is used as the flowing gas. Air is sent through a vinyl tube to the brass pipe by a fan (giving a flowrate of about 1000 liters/min). The wind velocity is controlled by adjusting the air blowing attachment. The wind velocity at the outlet is measured by a hot wire anemometer.

The temperature distribution and the corresponding refractive index variation in a gas lens is discussed in Sec. B.4.2. The system acts as a lens and focuses light beams through the tube.

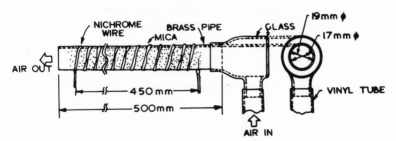

Fig. B.6. A gas lens of laminar flow type (after Aoki and Suzuki, 1966; reprinted by permission).

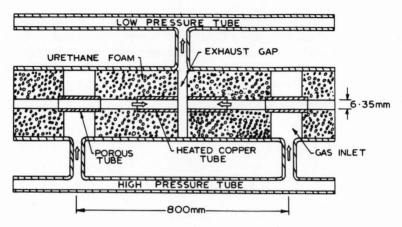

Fig. B.7. Section of a gas lens showing one counterflow lens (after Kaiser, 1970; reprinted by permission).

In the above experimental arrangement the effect due to gravity tends to distort the refractive index profile (Steier, 1965; Gloge, 1967). To avoid this the gas must be exhausted before the gravity effect becomes noticeable. This is the principle on which the gas lens shown in Fig. B.7 operates (Kaiser, 1970). The gas flow is in opposite directions in adjacent lens halves. The gas inlet is a porous pipe which guarantees uniform laminar flow in the end section of the path from inlet to outlet.

B.4.2.   Temperature and Refractive Index Variation in a Gas Lens

In a gas lens of laminar flow type the refractive index variation is constructed by the temperature gradients of a gas by setting up a cool gas flow through a warm metal pipe. For such a lens, if the effects of gravity, turbulence, and the influence of temperature on viscosity are neglected then the velocity distribution of the gas inside the lens is given by [see, for example, Jacob, 1949]:

$$V = 2V_m[1 - (r/a)^2] \tag{B.5}$$

where V_m is the mean velocity and a is the radius of the inner surface of the tube. The temperature distribution $T(r, z)$ in such a system is obtained by solving the following differential equation (Jacob, 1949):

$$\frac{\kappa}{\rho C_p}\left[\frac{\partial^2 T}{\partial r^2} + \frac{1}{r}\frac{\partial T}{\partial r}\right] = 2V_m\left[1 - \left(\frac{r}{a}\right)^2\right]\frac{\partial T}{\partial z} \tag{B.6}$$

where κ is the thermal conductivity, ρ is the density, and C_p is the specific heat of the gas. The solution of the above equation, subject to the boundary condition that

$$T(r = a, z) = T_s \tag{B.7}$$

for all values of z, has been extensively discussed (see, for example, Marcuse and Miller, 1964; Marcuse, 1965; Jacob, 1949) and is of the form

$$\frac{T_s - T(r, z)}{T_s - T_{00}} = -\sum_{n=0}^{\infty} \frac{2}{\beta_n \gamma_n} \exp(-\beta_n^2 \alpha z) \sum_{N=0,1,2} (B_{2N})_n \beta_n^{2N} \left(\frac{r}{a}\right)^{2N} \tag{B.8}$$

where

$$(B_0)_n = 1$$

$$(B_2)_n = -\frac{1}{4}$$

$$(B_{2N})_n = \frac{1}{4N^2} \left[\frac{1}{\beta_n^2} (B_{2N-4})_n - (B_{2N-2})_n \right]$$

$$\alpha = \frac{\kappa}{(2\rho C_p V_m a^2)}$$

$$\tag{B.9}$$

and

$$T_{00} = T(r = 0, z = 0)$$

is the input temperature and T_s is the wall temperature. The constants β_n and γ_n can be evaluated from detailed numerical calculations (Marcuse, 1965) and are tabulated in Table B.3. Once the temperature distribution is obtained, the refractive index variation can easily be calculated by using the following relation

$$[n(r, z) - 1]T(r, z) = [n_{00} - 1]T_{00} \tag{B.10}$$

where $n_{00} = n(r = 0, z = 0)$ is the refractive index corresponding to the temperature T_{00}.

If we substitute the expression for the temperature distribution from Eq. (B.8) in Eq. (B.10), we obtain the following refractive index variation

$$n(r, z) = n_0(z)[1 - a_2(z)r^2 + a_4(z)r^4 - a_6(z)r^6 + \cdots] \tag{B.11}$$

where

$$n_0(z) = 1 + (n_{00} - 1) - \frac{(n_{00} - 1)\theta_0}{T_{00}} - \frac{(n_{00} - 1)\theta_0}{T_{00}} \sum_{n=0}^{\infty} \frac{2}{\beta_n \gamma_n} \exp(-\beta_n^2 \alpha z)$$

$$\tag{B.12}$$

$$a_{2N}(z) = (-1)^N \frac{(n_{00}-1)\theta_0}{T_{00}n_0(z)a^{2N}} \sum_{n=0}^{\infty} (B_{2N})_n \frac{2}{\gamma_n} \beta^{2N-1} \exp(-\beta_n^2 \alpha z) \qquad \text{(B.13)}$$

and

$$\theta_0 = T_s - T_{00} \qquad \text{(B.14)}$$

The theory developed in Chapter 6 is valid for parabolic refractive index distribution. This essentially implies that we are assuming the temperature distribution to be given by

$$\frac{T_s - T(r, z)}{T_s - T_{00}} = -\sum_{n=0}^{\infty} \frac{2}{\beta_n \gamma_n} \exp(-\beta_n^2 \alpha z)\left[1 + (B_2)_n \beta_n^2 \left(\frac{r}{a}\right)^2\right] \qquad \text{(B.15)}$$

Clearly, the above equation does not satisfy the boundary condition given by Eq. (B.7) and for values of various parameters in a typical experimental arrangement, higher order terms in Eq. (B.8) make a significant contribution in determining the temperature profile. Thus, if we want to use the solutions given in Chapter 6, it is better to use the following approximate expression for the temperature distribution (Ghatak et al., 1973):

$$\frac{T_s - T(r, z)}{T_s - T_{00}} = \left[1 - \left(\frac{r}{a}\right)^2\right] \exp[-4\alpha z] \qquad \text{(B.16)}$$

Table B.3[a]

n	β_n	γ_n
0	2.70436	-0.50090
1	6.67903	0.37146
2	10.67338	-0.31826
3	14.67110	0.28646
4	18.66990	-0.26449
5	22.6691	0.24799
6	26.6686	-0.23491
7	30.6682	0.22485
8	34.6679	-0.21548
9	38.6676	0.20779
10	42.6667	-0.20108
11	46.6667	0.19516
12	50.6667	-0.18988
13	54.6667	0.18513
14	58.6667	-0.18083

[a] From Marcuse (1965).

Equation (B.16) satisfies Eq. (B.6) to the first order; it also satisfies the boundary conditions exactly and gives a temperature distribution which agrees with the exact calculations fairly well. Further, for the temperature distribution given by Eq. (B.16), $n_0(z)$ and $a_2(z)$ are given by

$$n_0(z) = 1 + (n_{00} - 1)\left\{ 1 - \frac{T_s - T_{00}}{T_{00}}[1 - \exp(-4\alpha z)]\right\} \qquad (B.17)$$

and

$$a_2(z) = \frac{2(n_{00} - 1)(T_s - T_{00})}{a^2 n_0(z) T_{00}} \exp(-4\alpha z) \qquad (B.18)$$

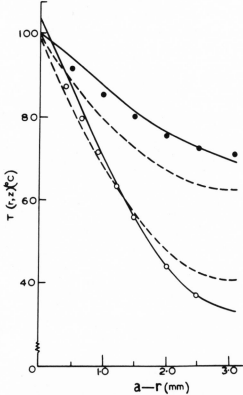

Fig. B.8. (○) and (●) represent the experimental data of Steier (1968) for the temperature profiles for a CO_2 gas lens at $z = 14.6$ and 4.8 cm, respectively. The wall temperature $T_s = 100°C$. The solid and dashed curves correspond to theoretical results using Eq. (B.8) and (B.16), respectively (after Ghatak et al., 1973; reprinted by permission).

Using the above expressions for $n_0(z)$ and $a_2(z)$ the solutions are given in Chapter 6.

Figure B.8 illustrates the calculated temperature distribution inside the gas lens and its comparison with experimental data of Steier (1968). The experiment uses CO_2 gas with a flowrate of 1.0 liters/min. The following values of various parameters have been used: $a = 0.3125$ cm, $\kappa/\rho C_p = 0.125$ cm^2/sec, and $V_m = 54.36$ cm/sec (as calculated from flowrate). The solid curves in Fig. B.8 correspond to the calculations using Eq. (B.8) where n goes up to 4 and N goes up to 5. The dashed curves in the figure correspond to Eq. (B.16). As expected the exact solution [Eq. (B.8)] agrees quite well with the experimental data. The agreement with the approximate solution [Eq. (B.16)] is also reasonably good and the difference with the experimental data is always less than 10%. However, the use of Eq. (B.15) leads to very poor agreement with the experimental data.

General Solution of Eq. (5.84)

We wish to find the general solution of the equation

$$\frac{\partial A_0^2}{\partial z} + r\left(\frac{1}{f}\frac{df}{dz}\right)\frac{\partial A_0^2}{\partial r} + 2\left(\frac{1}{f}\frac{df}{dz}\right)A_0^2 = 0 \tag{C.1}$$

where the function $f(z)$ is defined by the following equation:

$$\frac{1}{f}\frac{df}{dz} \equiv \beta \tag{C.2}$$

If we introduce a new variable

$$\xi = \frac{1}{f}$$

then

$$\begin{aligned}
\frac{\partial A_0^2}{\partial z} &= \frac{\partial A_0^2}{\partial \xi}\frac{d\xi}{dz} \\
&= -\frac{\partial A_0^2}{\partial \xi}\frac{1}{f^2}\frac{df}{dz}
\end{aligned} \tag{C.3}$$

Further, if we substitute for $\partial A_0^2/\partial z$ from Eq. (C.3) in Eq. (C.1) we get

$$-\xi\frac{\partial A_0^2}{\partial \xi} + r\frac{\partial A_0^2}{\partial r} + 2A_0^2 = 0 \tag{C.4}$$

Now we introduce two more variables

$$\eta = \frac{r}{r_0}\xi \tag{C.5}$$

and

$$\zeta = \frac{r}{r_0}\frac{1}{\xi} \tag{C.6}$$

Thus

$$\frac{\partial A_0^2}{\partial \xi} = \frac{\partial A_0^2}{\partial \eta}\frac{\partial \eta}{\partial \xi} + \frac{\partial A_0^2}{\partial \zeta}\frac{\partial \zeta}{\partial \xi}$$

$$= \frac{r}{r_0}\frac{\partial A_0^2}{\partial \eta} - \frac{r}{r_0}\frac{1}{\xi^2}\frac{\partial A_0^2}{\partial \zeta}$$

Similarly

$$\frac{\partial A_0^2}{\partial r} = \frac{\xi}{r_0}\frac{\partial A_0^2}{\partial \eta} + \frac{1}{r_0\xi}\frac{\partial A_0^2}{\partial \zeta}$$

Substituting the above results in Eq. (C.4) we obtain, after some simplifications,

$$\zeta\frac{\partial A_0^2}{\partial \zeta} + A_0^2 = 0$$

or

$$\frac{\partial}{\partial \zeta}(\zeta A_0^2) = 0$$

or $\zeta A_0^2 =$ an arbitrary function of η. But

$$\zeta = \frac{r}{r_0}\frac{1}{\xi} = \frac{\eta}{\xi^2}$$

Therefore $A_0^2 = \xi^2 \times$ another arbitrary function of η. Thus when S is given by Eq. (5.83), the general solution of Eq. (5.76) is of the form

$$A_0^2 = \frac{E_0^2}{f^2}\mathscr{F}\left(\frac{r}{r_0 f(z)}\right) \tag{C.7}$$

where $\mathscr{F}$ is an arbitrary function of its argument.

Derivation of the Ray Equation

In this appendix we give a simple derivation of the ray equation from Fermat's principle. In Chapter 9 we had obtained the optical Lagrangian equations from Fermat's principle. These are [see Eqs. (9.6) and (9.7)]

$$\frac{d}{dz}\left(\frac{\partial L}{\partial \dot{x}}\right) = \frac{\partial L}{\partial x} \tag{D.1a}$$

$$\frac{d}{dz}\left(\frac{\partial L}{\partial \dot{y}}\right) = \frac{\partial L}{\partial y} \tag{D.1b}$$

where

$$L = n(x, y, z)(1 + \dot{x}^2 + \dot{y}^2)^{1/2} \tag{D.2}$$

where the dots represent derivatives with respect to z. If we substitute for L from Eq. (D.2) in Eq. (D.1) we get

$$\frac{d}{dz}\left[n\frac{\dot{x}}{(1+\dot{x}^2+\dot{y}^2)^{1/2}}\right] = (1+\dot{x}^2+\dot{y}^2)^{1/2}\frac{\partial n}{\partial x} \tag{D.3}$$

Since

$$ds = (1 + \dot{x}^2 + \dot{y}^2)^{1/2}\, dz$$

we get

$$\frac{1}{(1+\dot{x}^2+\dot{y}^2)^{1/2}}\frac{d}{dz}=\frac{d}{ds} \tag{D.4}$$

Thus

$$\frac{d}{ds}\left(n\frac{dx}{ds}\right)=\frac{\partial n}{\partial x} \tag{D.5}$$

Similarly for the y component we obtain

$$\frac{d}{ds}\left(n\frac{dy}{ds}\right)=\frac{\partial n}{\partial y} \tag{D.6}$$

Equations (D.5) and (D.6) represent the x and y components of the ray equation. The z component can be obtained from these two equations meaning thereby that only two components of the ray equation are independent.

Now,

$$\frac{d}{ds}\left(n\frac{dz}{ds}\right)=\frac{dn}{ds}\frac{dz}{ds}+n\frac{d^2z}{ds^2} \tag{D.7}$$

Since

$$\frac{dz}{ds}=\left[1-\left(\frac{dx}{ds}\right)^2-\left(\frac{dy}{ds}\right)^2\right]^{1/2} \tag{D.8}$$

we get

$$n\frac{d^2z}{ds^2}=\left[1-\left(\frac{dx}{ds}\right)^2-\left(\frac{dy}{ds}\right)^2\right]^{-1/2}\left\{\frac{dn}{ds}\left[\left(\frac{dx}{ds}\right)^2+\left(\frac{dy}{ds}\right)^2\right]-\frac{\partial n}{\partial x}\frac{dx}{ds}-\frac{\partial n}{\partial y}\frac{dy}{ds}\right\} \tag{D.9}$$

where we have used [see Eqs. (D.5) and (D.6)]:

$$n\frac{d^2x}{ds^2}=\frac{\partial n}{\partial x}-\frac{dn}{ds}\frac{dx}{ds} \tag{D.10}$$

and

$$n\frac{d^2y}{ds^2}=\frac{\partial n}{\partial y}-\frac{dn}{ds}\frac{dy}{ds} \tag{D.11}$$

Substituting for $n\,d^2z/ds^2$ from Eq. (D.9) in Eq. (D.7) we get

$$\frac{d}{ds}\left(n\frac{dz}{ds}\right)=\frac{dn}{dz}-\frac{\partial n}{\partial x}\frac{dx}{dz}-\frac{\partial n}{\partial y}\frac{dy}{dz} \tag{D.12}$$

Since

$$\frac{dn}{dz} = \frac{\partial n}{\partial z} + \frac{\partial n}{\partial x}\frac{dx}{dz} + \frac{\partial n}{\partial y}\frac{dy}{dz} \tag{D.13}$$

we get

$$\frac{d}{ds}\left(n\frac{dz}{ds}\right) = \frac{\partial n}{\partial z} \tag{D.14}$$

which is the z component of the ray equation.

References

Akhmanov, S. A., Sukhorukov, A. P., and Khokhlov, R. V. (1966). Self-Focusing and Self-Trapping of Intense Light Beams in a Non Linear Medium, *Sov. Phys. JETP* **23**: 1025.

Akhmanov, S. A., Sukhorukov, A. P., and Khokhlov, R. V. (1968). Self-Focusing and Diffraction of Light in a Nonlinear Medium, *Sov. Phys. Uspekhi* **10**: 609.

Allan, W. B. (1973). *Fibre Optics: Theory and Practice*, Plenum, London.

Aoki, Y., and Suzuki, M. (1967). Imaging Properties of a Gas Lens, *IEEE Trans. Microwave Theory Tech.* **MTT-15**: 2.

Ballman, A. A., Levinstein, H. J., Capio, C. P., and Brown, H. (1967). Curie Temperature and Birefringence Variation in Ferroelectric Lithium Metatantalate as a Function of Melt Stoichiometry, *J. Am. Ceram. Soc.* **50**: 657.

Barnoski, M. (ed) (1974). *Introduction to Integrated Optics*, Plenum, New York.

Barns, R. L., and Carruthers, J. R. (1970). Lithium Tantalate Single Crystal Stoichiometry, *J. Appl. Cryst.* **3**: 395.

Bergman, J. G., Ashkin, A., Ballman, A. A., Dziedzic, J. M., Levinstein, H. J., and Smith, R. G. (1968). Curie Temperature, Birefringence, and Phase Matching Temperature Variations in LiNbO₃ as a Function of Melt Stoichiometry, *Appl. Phys. Lett.* **12**: 92.

Born, M., and Wolf, E. (1970). *Principles of Optics*, Fourth Edition, Pergamon, London.

Bouillie, R., Cozannet, A., Steiner, K. H., and Treheux, M. (1974). Ray Delay in Gradient Waveguides with Arbitrary Symmetric Refractive Profile, *Appl. Opt.* **13**: 1045.

Buchdahl, H. A. (1968). *Optical Aberration Coefficients*, Dover, New York.

Buchdahl, H. A. (1970). *An Introduction to Hamiltonian Optics*, Cambridge Univ. Press, Cambridge, England.

Carruthers, J. R., Peterson, G., Grasso, M., and Bridenbaugh, P. M. (1971). Nonstoichiometry and Crystal Growth of Lithium Niobate, *J. Appl. Phys.* **42**: 1846.

Casperson, L. W. (1973). Gaussian Light Beams in Inhomogeneous Media, *Appl. Opt.* **12**: 2434.

Clarricoats, P. J. B. (1977). Theory of Optical Fiber Waveguides: A Review, in *Progress in Optics*, E. Wolf, Vol. XIV, in press.

Conwell, E. (1973). Modes in Optical Waveguides Formed by Diffusion, *Appl. Phys. Lett.* **23**: 328.

Grank, J. (1956). *Mathematics of Diffusion*, Oxford Univ. Press, Oxford, England.

Dakin, J. P., Gambling, W. A., Matsumura, H., Payne, D. N., and Sunak, H. R. D. (1973). Theory of Dispersion in Lossless Multimode Optical Fibers, *Opt. Commun.*, **7**: 1.

Gambling, W. A., and Matsumura, H. (1973). Pulse Dispersion in a Lenslike Medium, *Opt. Electron.* **5**: 429.

Gambling, W. A., Payne, D. N., and Matsumura, H. (1972). Dispersion in Low-Loss Liquid Core Optical Fibers, *Electron. Lett.* **8**: 568.

Gambling, W. A., Payne, D. N., Matsumura, H., and Medlicott, M. (1974). Optical Fibers and the Goos–Hanchen Shift, *Electron. Lett.* **10**: 99.

Garrett, C. G. B., and McCumber, D. E. (1970). Propagation of a Gaussian Light Pulse Through an Anomalous Dispersion Medium, *Phys. Rev. A* **1**: 305.

Gedeon, A. (1974). Comparison Between Rigorous Theory and WKB-Analysis of Modes in Graded Index Waveguides, *Opt. Commun.* **12**: 329.

Geshiro, M., Ootaka, M., Matsuhara, M., and Kumagai, N. (1974). Analysis of Wave Modes in Slab Waveguides with Truncated Parabolic Index, *IEEE J. Quant. Electron. (Corresp.)* **QE-10**: 647.

Ghatak, A. K., and Kraus, L. A. (1974). Propagation of Waves in a Medium Varying Transverse to the Direction of Propagation, *IEEE J. Quant. Electron. (Corresp.)* **QE-10**: 465.

Ghatak, A. K., and Lokanathan, S. (1975). *Quantum Mechanics*, Macmillan of India Ltd., New Delhi.

Ghatak, A. K., and Thyagarajan, K. (1975). Ray and Energy Propagation in Graded Index Media, *J. Opt. Soc. Am.* **65**: 169.

Ghatak, A. K., Singh, K. Malik, D. P. S., and Sodha, M. S. (1972). Resolution by SELFOC Lens, *Opt. Acta* **19**: 681.

Ghatak, A. K., Malik, D. P. S., and Goyal, I. C. (1973). Electromagnetic Wave Propagation through a Gas Lens, *Opt. Acta* **20**: 303.

Ghatak, A. K., Goyal, I. C. and Kumar, Arun (1975). Propagation of Gaussian Pulse through an Optical Fiber: Applicability of Geometrical Optics, *Appl. Opt.* **14**: 2330.

Ghatak, A. K., Goyal, I. C., and Sharma, A. (1976). Mode Conversion in Optical Waveguides, *Opt. Quant. Electron.* **8**: 399.

Gloge, D. (1967). Deformation of Gas Lens by Gravity, *Bell Syst. Tech. J.* **46**: 357.

Gloge, D. (1970). Optical Waveguide Transmission, *Proc. IEEE* **58**: 1513.

Gloge, D. (1971a). Weakly Guiding Fibers, *Appl. Opt.* **10**: 2252.

Gloge, D. (1971b). Dispersion in Weakly Guiding Fibers, *Appl. Opt.* **10**: 2442.

Gloge, D. (1974). Optical Fibers for Communication, *Appl. Opt.* **13**: 249.

Gloge, D., and Marcatili, E. A. J. (1973). Multimode Theory of Graded-core Fibers, *Bell Syst. Tech. J.* **52**: 1563.

Gloge, D., Chinnock, E., and Lee, T. P. (1974). GaAs Twin Laser Set up to Measure Mode and Material Dispersion in Optical Fibers, *Appl. Opt.* **13**: 261.

Goell, J. E., and Standley, R. D. (1969). Sputtered Glass Waveguide for Integrated Optical Circuits, *Bell Syst. Tech. J.* **48**: 3445.

Goldstein, H. (1950). *Classical Mechanics*, Addison–Wesley, Reading, Massachusetts.

Goodman, J. W. (1968). *An Introduction to Fourier Optics*, McGraw-Hill, New York.

Goyal, I. C., Sodha, M. S., and Ghatak, A. K. (1973). Propagation of Electromagnetic Waves in Medium with Random Radial Dielectric Constant Gradient, *J. Opt. Soc. Am.* **63**: 940.

Goyal, I. C., Kumar, Arun, and Ghatak, A. K. (1976). Calculation of Bandwidth of Optical Fibers from Experiments on Dispersion Measurement, *Opt. Quant. Electron.* **8**: 80.

Goyal, I. C., Kumar, A., and Ghatak, A. K. (1976). Calculation of Bandwidth of Optical Fibers from Experiments on Dispersion Measurement, *Opt. Quant. Electron.* **8**: 80.

Gupta, A., Thyagarajan, K., Goyal, I. C., and Ghatak, A. K. (1976). Theory of Fifth Order Aberration in Graded Index Media, *J. Opt. Soc. of Am.* December, 1976, in press.

Harris, J. H., Shubert, R., and Polky, J. N. (1970). Beam Coupling to Films, *J. Opt. Soc. Am.* **60**: 1007.

Irving, J., and Mullineux, I. (1959). Mathematics in Physics and Engineering, Academic, New York.

Izawa, T., and Nakagome, H. (1972). Optical Waveguide Formed by Electrically Induced Migration of Ions in Glass Plates, *Appl. Phys. Lett.* **21**: 584.

Jacob, M. (1949). *Heat Transfer* Vol. 1, Wiley, New York.

Kaiser, P. (1968a). Measured Beam Deformations in a Guide Made of Tubular Gas Lenses, *Bell Syst. Tech. J.* **47**: 179.

Kaiser, P. (1968b). The Stream Guide, a Simple, Low- Loss Optical Guiding Medium, *Bell Syst. Tech. J.* **47**: 761.

Kaiser, P. (1970). An Improved Thermal Gas Lens for Optical Beam Waveguides, *Bell Syst. Tech. J.* **49**: 137.

Kaminow, I. P., and Carruthers, J. R. (1973). Optical Waveguiding Layers in $LiNbO_3$ and $LiTaO_3$, *Appl. Phys. Lett.* **22**: 326.

Kaminow, I. P., Mammel, W. L., and Weber, H. P. (1974). Metal-clad Optical Waveguides: Analytical and Experimental Study, *Appl. Opt.* **13**: 396.

Kapany, N. S. (1967). *Fiber Optics*, Academic, New York.

Kapany, N. S., and Burke, J. J. (1972). *Optical Waveguides*, Academic, New York.

Kapron, F. P., and Keck, D. B. (1971). Pulse Transmission through a Dielectric Optical Waveguide, *Appl. Opt.* **10**: 1519.

Kawakami, S., and Nishizawa, J. (1968). An Optical Waveguide with the Optimum Distribution of Refractive Index with Reference to Waveform Distortion, *IEEE Trans. Microwave Theory Tech.* **MTT-16**: 814.

Khular, E., and Ghatak, A. K. (1976). Pulse Dispersion in Planar Waveguides having Exponential Variation of Refractive Index, *Optica Acta*, December, 1976, in press.

Kita, H., and Uchida, T. (1969). Light-Focusing Glass Fiber and Rod, SPIE Fiber Optics Seminar, Soc. of Photo-Optical Instrumentation Engineers, Redondo Beach, California.

Kita, H., Kitano, I., Uchida, T., and Furukawa, M. (1971). Light-Focusing Glass Fibers and Rods, *J. Amer. Ceram. Soc.* **54**: 321.

Kitano, I., Koizumi, K., Matsumura, H., Uchida, T. and Furukawa, M. (1970). A Light-Focusing Fiber Guide Prepared by Ion-Exchange Techniques, *Japan Soc. Appl. Phys. (Suppl.)*, **39**: 63.

Kitano, T., Matsumura, H., Furukawa, M., and Kitano, I. (1973). Measurement of Fourth-Order Aberration in a Lenslike Medium, *IEEE J. Quant. Electron.* **QE-9**: 967.

Kogelnik, H. (1975). An Introduction to Integrated Optics, *IEEE Trans. Microwave Theory Tech.* **MTT-23**: 1.

Koizumi, K., Ikeda, Y., Kitano, I., Furukawa, M., and Sumimoto, T. (1974). New Light-Focusing Fibers Made by a Continuous Process, *Appl. Opt.* **13**: 255.

Kumar, A., Thyagarajan, K., and Ghatak, A. K. (1974). Modes in Inhomogeneous Slab-Waveguides, *IEEE J. Quant. Electron.* **QE-10**: 902.

Kurtz, C. N., and Streifer, W. (1969a). Guided Waves in Inhomogeneous Focusing Media, Part I: Formulation, Solution for Quadratic Inhomogeneity, *IEEE Trans. Microwave Theory Tech.* **MTT-17**: 11.

Kurtz, C. N., and Streifer, W. (1969b). Guided Waves in Inhomogeneous Focusing Media, Part III: Asymptotic Solution for General Weak Inhomogeneity, *IEEE Trans. Microwave Theory Tech.* **MTT-17**: 250.

Lisitsa, M. P., Gudymenko, L. F., Malinko, V. N., and Terekhova, S. F. (1969). Dispersion of the Refractive Indices and Birefringence of CdS_xSe_{1-x} Single Crystals, *Phys. Stat. Sol.* **31**: 389.

Luneburg, R. K. (1964). *Mathematical Theory of Optics*, Univ. California Press, Berkeley.

Marcuse, D. (1965). Properties of Periodic Gas Lenses, *Bell Syst. Tech. J.* **44**: 2083.

Marcuse, D. (1969). Mode Conversion Caused by Surface Imperfections of a Dielectric Slab Waveguide, *Bell Syst. Tech. J.* **48**: 3187.

Marcuse, D. (1972). *Light Transmission Optics*, Van Nostrand–Reinhold, Princeton, New Jersey.

Marcuse, D. (1973a). The Effect of the ∇n^2 Term on the Modes of an Optical Square Law Medium, *IEEE J. Quant. Electron.* (*Corresp.*) **QE-9**: 958.

Marcuse, D. (1973b). TE Modes of Graded-Index Slab-Waveguides, *IEEE J. Quant. Electron* **QE-9**: 1000.

Marcuse, D. (1973c). Coupling Coefficients for Imperfect Asymmetric Slab Waveguides, *Bell Syst. Tech. J.* **52**: 63.

Marcuse, D. (ed) (1973d). *Integrated Optics*, IEEE, New York.

Marcuse, D. (1973e). The Impulse Response of an Optical Fiber With Parabolic Index Profile, *Bell Syst. Tech. J.* **52**: 1169.

Marcuse, D. (1974). *Theory of Dielectric Optical Waveguides*, Academic, New York.

Marcuse, D., and Miller, S. E. (1964). Analysis of a Tubular Gas Lens, *Bell Syst. Tech. J.* **43**: 1759.

Margenau, H., and Murphy, G. M. (1956). The Mathematics of Physics and Chemistry, Van Nostrand–Reinhold, Princeton, New Jersey.

Matsuhara, M. (1973). Analysis of TEM Modes in Dielectric Waveguides, by a Variational Method, *J. Opt. Soc. Am.* **63**: 1514.

Maurer, R. D. (1973). Glass Fibers for Optical Communications, *Proc. IEEE* **61**: 452.

Maurer, R. D. (1974). In *Introduction to Integrated Optics* Barnoski, M. K. (ed), Chapter 8, Plenum, New York.

Miller, S. E. (1965). Light Propagation in Generalized Lenslike Media, *Bell Syst. Tech. J.* **44**: 2017.

Miller, S. E., Marcatili, E. A. J., and Li, T. (1973). Research Toward Optical-Fiber Transmission Systems Part I: The Transmission Medium; Part II: Devices and Systems Considerations, *Proc. IEEE* **61**: 1703.

Montagrino, L. (1968). Ray Tracing in Inhomogeneous Media *J. Opt. Soc. Am.* **58**: 1667.

Moore, D. T. (1971). Design of Singlets with Continuously Varying Indices of Refraction, *J. Opt. Soc. Am.* **61**: 886.

Morse, P. M. and Feshbach, H. (1953). *Methods of Theoretical Physics*, Vol. II, McGraw-Hill, New York.

Murphy, G. M. (1960). *Ordinary Differential Equations*, Van Nostrand–Reinhold, Princeton, New Jersey.

Nelson, D. F., and Mckenna, J. (1967). Electromagnetic Modes of Anisotropic Dielectric Waveguide at $p-n$ Junction, *J. Appl. Phys.* **38**: 4057.

Ohtsuka, Y. (1973). Light-focusing Plastic Rod Prepared from Diallyl Isophthalate-methyl Methacrylate Copolymerization, *Appl. Phys. Lett.* **23**: 247.

Otto, A., and Sohler, W. (1971). Modification of the Total Reflection Modes in a Dielectric Film by One Metal Boundary, *Opt. Commun.* **3**: 254.

Pearson, A. D., French, W. G., and Rawson, E. G. (1969). Preparation of a Light-Focusing Glass Rod by Ion Exchange Techniques, *Appl. Phys. Lett.* **15**: 76.

Rawson, E. G., Herriott, D. R., and Mckenna, J. (1970). Analysis of Refractive Index Distributions in Cylindrical Graded Index Glass Rods (GRIN Rods) used as Image Relays, *Appl. Opt.* **9**: 753.

Reinhart, F. K., Nelson, D. F., and Mckenna, J. (1969). Electro-optic and Waveguide Properties of Reverse-Biased Gallium Phosphide p–n Junctions, *Phys. Rev.* **177**: 1208.

Sands, P. J. (1970). Aberrations of Inhomogeneous Lenses, *J. Opt. Soc. Am.* **60**: 1436.

Schiff, L. I. (1955). *Quantum Mechanics*, McGraw-Hill, New York.

Schiff, L. I. (1968). *Quantum Mechanics*, Third Edition, McGraw-Hill, New York.

Schlesinger, S. P., Diament, P., and Vigants, A. (1960). On Higher Order Hybrid Modes of Dielectric Cylinders, *IEEE Trans. Microwave Theory Tech.* **MTT-8**: 252.

Smithgall, D. H. and Dabby, F. W. (1973). Graded-Index Planar Dielectric Waveguides, *IEEE J. Quant. Electron.* **QE-9**: 1023.

Smirnov, A. D. (1960). *Tables of Airy Functions and Confluent Hypergeometric Functions*, Pergamon, London.

Snyder, A. W. (1969). Asymptotic Expressions for Eigenfunctions and Eigenvalues of Dielectric or Optical Waveguide, *IEEE Trans. Microwave Theory Tech.* **MTT-17**: 1130.

Sodha, M. S., Ghatak, A. K. and Malik, D. P. S. (1971a). Electromagnetic Wave Propagation in a Radially and Axially Nonuniform Media, Geometrical Optics Approximation, *J. Opt. Soc. Am.* **61**: 1492.

Sodha, M. S., Ghatak, A. K., and Malik, D. P. S. (1971b). Electromagnetic Wave Propagation in a Radially and Axially Nonuniform Media: Optics of Selfoc Fibers and Rods, Wave Optics Considerations, *J. Phys. D. Appl. Phys.* **4**: 1887.

Sodha, M. S., Ghatak, A. K., and Goyal, I. C. (1972a). Series Solution for Electromagnetic Wave Propagation in Radially and Axially Nonuniform Media: Geometrical Optics Approximation, *J. Opt. Soc. Am.* **62**: 963.

Sodha, M. S., Ghatak, A. K., Tewari, D. P., and Dubey, P. K. (1972b). Focusing of Waves in Ducts, *Radio Sci.* **7**: 1005.

Sodha, M. S., Chakravarti, A. K., and Gautama, G. D. (1973). Propagation of Optical Pulses through Cladded Fibers: Modified Theory, *Appl. Opt.* **12**: 2482.

Sodha, M. S., Ghatak, A. K., and Tripathi, V. K. (1974). *Self-Focusing of Laser Beams in Dielectrics Plasmas and Semiconductors*, Tata McGraw-Hill, New Delhi.

Steier, W. H. (1965). Measurements on a Thermal Gradient Gas Lens, *IEEE Trans. Microwave Theory Tech.* **MTT-13**: 740.

Steier, W. H. (1968). Optical Shuttle Pulse Measurements of Gas Lenses, *Appl. Opt.* **7**: 2295.

Steward, G. C. (1928). *The Symmetrical Optical System*, Cambridge Univ. Press, Cambridge, England.

Taylor, H. F., Martin, W. E., Hall, D. B., and Smiley, V. N. (1972). Fabrication of Single Crystal Semiconductor Optical Waveguides, *Appl. Phys. Lett.* **21**: 95.

Thyagarajan, K., and Ghatak, A. K. (1974). Perturbation Theory for Studying the Effect of the $\nabla \varepsilon$ Term in Lens-Like Media, *Opt. Commun.* **11**: 417.

Thyagarajan, K., and Ghatak, A. K. (1976). Hamiltonian Theory of Third-Order Aberrations of Inhomogeneous Lenses, *Optik* **44**: 329.

Tien, P. K. (1971). Light Waves in Thin Films and Integrated Optics, *Appl. Opt.* **10**: 2395.

Tien, P. K., and Ulrich, R. (1970). Theory of Prism-Film Coupler and Thin-Film Light Guides, *J. Opt. Soc. Am.* **60**: 1325.

Tien, P. K., Ulrich, R., and Martin, R. J. (1969). Modes of Propagating Light Waves in Thin Deposited Semiconductor Films, *Appl. Phys. Lett.* **14**: No. 9, 291.

Tien, P. K., Smolinsky, G., and Martin, R. J. (1972). Thin Organosilicon Films for Integrated Optics, *Appl. Opt.* **11**: 637.

Uchida, T., Furukawa, M., Kitano, I., Koizumi, K., and Matsumura, H. (1970). Optical Characteristics of a Light-Focusing Fiber Guide and Its Applications, *IEEE J. Quant. Electron.* **QE-6**: 606.

Ulrich, R. (1970). Theory of the Prism-Film Coupler by Plane-Wave Analysis, *J. Opt. Soc. Am.* **60**: 1337.

Woodbury, H. H., and Hall, R. B. (1967). Diffusion of the Chalogens in the II–VI Compounds, *Phys. Rev.* **157**: 641.

Yamada, R., and Y. Inabe, (1974). Guided Waves in an Optical Square Low Medium, *J. Opt. Soc. Am.* **64**: 964.

Zachos, T. H., and Ripper, J. E. (1969). Resonant Modes of GaAs Junction Lasers, *IEEE J. Quant. Electron.* **QE-5**: 29.

Author Index

Subject Index